Direct Digital Control of Building Systems

Direct Digital Control of Building Systems

Theory and Practice

H. Michael Newman
Manager
Facilities Engineering Computer Section
Cornell University

With editorial assistance from:
M. D. Morris, P.E.

A WILEY-INTERSCIENCE PUBLICATION
John Wiley & Sons, Inc.
NEW YORK / CHICHESTER / BRISBANE / TORONTO / SINGAPORE

This text is printed on acid-free paper.

Library of Congress Cataloging in Publication Data:

Newman, H. Michael, 1942–
Direct digital control of building systems : theory and practice / by H. Michael Newman.
p. cm. — (Wiley series of practical construction guides.)
Includes bibliographical references and index.
ISBN 0-471-51696-1 (cloth : acid free paper)
1. Heating—Automatic control. 2. Ventilation—Automatic control. 3. Air conditioning—Automatic control. 4. Digital control systems. I. Title. II. Series.
TH7466.5.N48 1994
697—dc20 93-7907
CIP

Printed in the United States of America

10 9 8 7 6 5 4 3 2 1

To my parents, who didn't mind that I decided
to become an engineer instead of a lawyer
and
To my son, Kevin, from whom I learn new things every day

Preface: Why DDC?

We live in an *analog* world! Temperature, pressure, fluid flow, and virtually all the other physical quantities that must be controlled in heating, ventilating, air-conditioning, and refrigerating systems are continuously variable, analog quantities. What then is responsible for the increasing dominance of *digital* controls in the world of HVAC&R, a world formerly ruled by pneumatic and electronic analog controls?

Of the numerous justifications for using digital controls in place of the more traditional technologies, three stand out:

1. Digital controls cost less.

Due to their high reliability, digital controls often cost less both initially and in terms of their life cycle cost. This has not always been the case, but the dramatic decline in the cost of microprocessors, computer memory, and other electronic components during the last decade has been felt by all computer manufacturers including those who use computers for monitoring and control. This, along with the increasing sophistication of the controls vendors' branch office personnel responsible for developing the software for specific applications, has led to significant reductions in the installed cost of DDC versus the older systems.

Indeed, as vendors have geared up for dealing with the technical challenge posed by computer-based technology, they have begun to discover some of the appeal of computers long appreciated by computer specialists. One of these is that if you have a program that *works* you can reuse it, in whole or in part, the next time you need to solve a similar programming problem. Since HVAC control requirements are often fairly similar from one system to the next, the vendors have discovered the joy of being able to transport successful code from one digital controller to another often with substantial savings in time, effort, and cash.

2. Digital controls can perform the same task as older controls only "better."

Computers process numbers, which can be stored in a computer's memory indefinitely without changing. Since a digital controller's operating parameters (e.g., control loop setpoints, gains, operating times, calibration factors, etc.) are stored as numbers, they never change! This is not the case with analog controllers where the operating parameters are stored as a particular amount of spring tension, airflow through an orifice, or voltage across a sensing circuit. As a result, digital controls do not drift or "wander" over time. (It *is* true that the analog input and output devices may require occasional recalibration, but not the controllers themselves.)

Digital controls also have the inherent flexibility to adapt to altered or refined control requirements, often without having to add more hardware. Once the various sensor inputs and actuator outputs have been connected, the designer can establish virtually any conceivable relationship between them. The same sensor, for example, can be used as an input to many different processes. If an actuator requires hysteresis correction or needs to be "ramped" to its operating position, this can often be accomplished simply by invoking software capabilities that are "built in" to the machine.

3. Digital controls can perform tasks that other controls cannot do at all.

Computers, for example, can store and retrieve information for later review. It is thus possible to accumulate historical data on the operation of building systems that can be used to spot mechanical problems before they become serious or can help to improve energy efficiency. Along the same line, digital controllers can serve as meters for various utilities and can be programmed to display the results in whatever units are desired.

Because moving information from one place to another is an essential part of a computer's operation, digital controllers can easily be made to communicate with other computers. Thus a controller which has been installed primarily to perform *control* can often provide remote *monitoring* capability almost as a fringe benefit.

The ability to peer into the heart of the controller from afar also means that its functions can be tuned or adjusted remotely. (This is good news for those who don't like climbing stepladders!) Coupled with graphic plotting or trending capability, a technician can often diagnose and sometimes correct a problem without leaving his or her office.

In short, computer-based monitoring and control systems offer a plentiful range of application possibilities that can open the door to energy savings, reduced maintenance costs, and greater comfort. The trick is that they must be correctly and appropriately applied. It is the goal of this book to give the reader the tools and insights necessary to do just that.

HOW TO USE THIS BOOK

Most of the chapters in this book can be treated as "stand-alone" entities and can therefore be tackled in any order. The only real exceptions to this are the chapters dealing with data communication and BACnet. Unless you are already familiar with data communication, you should definitely read these chapters in sequence. I have also tried to make access to further information easy by providing references to the various standards and other documents that make up so much of the foundation of DDC technology. In addition to a fairly extensive glossary and index, I have provided the addresses of most of the current suppliers of DDC hardware and software who will be happy to supply you with the specifics of their products.

Ithaca, New York H. MICHAEL NEWMAN

Acknowledgments

From the outset I have had the good fortune to enjoy the support, advice, and encouragement of many people in the preparation of this book. Without these colleagues, friends, and family members this book would simply never have come into being. I would therefore like to specifically thank:

Steve Bushby of NIST who has been instrumental in the development of BACnet and who made many excellent suggestions for improving this book; Dave Fisher of American Auto-Matrix who has also worked tirelessly for rationality in computer communications; Bert Murphy of Universal Software who helped me round out the list of DDC suppliers; and Clay Nesler of Johnson Controls who made numerous suggestions and with whom I have had the privilege of teaching ASHRAE's Professional Development Seminar on DDC for the past several years.

Many of my Cornell colleagues have also been of great help. Among them are Bruce Bush, Chuck Eckert, John Hall, and Ken Potter, of our Controls Shop and EMCS Operations staff, all of whom have been generous with their knowledge, insights, and suggestions concerning the practical side of controls. Among the engineering staff who were particularly encouraging to me in my struggle were Steve Little, who knows something about everything, Karl Schmid, our former director, Wendy Fuller, and Daisy Kirkpatrick. Most particularly I want to thank Joel Bender with whom I have worked since the early 1980s. His software wizardry has made possible the implementation of so many innovative EMCS capabilities that they almost defy enumeration.

Speaking of innovation, I am happy to thank Peter Macdonald for his humorous cartoon-like illustrations that I have used to set the tone for most of the chapters. There is, believe it or not, an occasional element of humor in trying to get computers to do something useful!

Many friends have likewise provided support and encouragement, if only by being sure to ask "OK, Newman, where is it?" Among these I would like to mention Tim Gardner, Pete Hoyt, Pete Kortman, Mel Richards, and Karl J. "Kopfjäger" Schoellkopf.

Thanks also go to my sisters Dorci, Barbara, and Wendy, and, of course, my Mother and Father. Although my Dad is no longer with us, and would not have understood a word of this text anyway, I know he would have wanted to be able to tell his friends about "his son, the author!"

Finally, my special thanks and gratitude go to my friend, mentor, and erstwhile editor, M. Dan Morris, without whose initiative, patience, and occasional gentle prodding, this book would have forever remained nothing more than a future project.

H. M. N.

Contents

Preface: Why DDC? **vii**

1 Introduction **1**

Summary / 1
1.1 Historical Perspective / 2
1.2 Automatic Feedback and Early Building Controls / 3
1.3 Manual Central Control Systems / 4
1.4 Computerized Manual Systems / 6
1.5 The Field Panel Interface / 6
1.6 Stand-alone DDC / 7
1.7 Distributed DDC / 8
References / 9

2 Sensors and Actuators **11**

Summary / 11
2.1 Sensors / 12
2.1.1 Temperature Sensors / 13
2.1.1.1 Resistance Temperature Detectors / 14
2.1.1.2 Thermistors / 18
2.1.1.3 Thermocouples / 21
2.1.1.4 Integrated Circuit Temperature Sensors / 23
2.1.1.5 Temperature Sensor Summary / 23
2.1.2 Pressure Sensors / 24
2.1.2.1 Capacitive Transducers / 25
2.1.2.2 Inductive Transducers / 25
2.1.2.3 Strain Gauge Transducers / 26
2.1.2.4 Piezoelectric Transducers / 27

2.1.2.5 Potentiometric Transducers / 27
2.1.2.6 Pressure Sensor Summary / 27
2.1.3 Velocity and Flow Sensors / 29
2.1.3.1 Pitot Tubes / 31
2.1.3.2 Orifice Plates / 33
2.1.3.3 Venturi Tubes / 33
2.1.3.4 Hot-Wire Anemometers / 34
2.1.3.5 Turbine Flow Meters / 36
2.1.3.6 Vortex Shedding Meters / 36
2.1.3.7 Ultrasonic Flow Meters / 37
2.1.4 Humidity Sensors / 39
2.1.5 Indoor Air Quality (IAQ) Sensors / 40
2.2 Sensor Signal Conditioning / 41
2.2.1 Multiplexing and Analog/Digital Conversion / 42
2.3 Sensors of the Future / 45
2.4 Actuators and Other Output Devices / 46
2.4.1 Binary Output Devices / 46
2.4.2 Analog Output Devices / 47
2.4.2.1 Digital/Analog Conversion / 48
2.4.2.2 Analog Output Transducers: *V/P, I/P* / 48
2.4.2.3 Pulse Width Modulation / 50
References / 51

3 DDC Hardware **53**

Summary / 53
3.1 Fundamentals / 54
3.2 Data Representation / 54
3.3 Microprocessors/Microcomputers/Microcontrollers / 58
3.4 Microprocessor Programming / 60
3.5 DDC Packaging / 63
3.5.1 General-Purpose DDCs / 63
3.5.2 Application Specific or Unitary Controllers / 65
3.5.3 "Smart" Sensors and Actuators / 66
Bibliography / 67

4 DDC Software **69**

Summary / 69
4.1 Software Concepts / 70
4.2 Operating System Software / 70
4.2.1 Workstation Operating Systems / 73
4.2.2 DDC Field Panel Operating Systems / 73

4.3 Utility Software / 74
4.3.1 Workstation Utility Software / 74
4.3.2 DDC Field Panel Utility Software / 75
4.4 Application Software / 76
4.4.1 Energy Management Software / 76
4.4.2 Metering and Monitoring Software / 79
4.4.3 Additional Application Software / 80
4.5 Configuring DDCs: Overview / 81
4.6 Configuration Methods / 82
4.6.1 Database Entry / 82
4.6.2 DDC Programming / 83
4.6.2.1 DDC Programming Languages / 84
4.6.2.2 Translators and Compilers / 85
4.6.3 Graphical Programming and Other Techniques / 87
4.7 Backup / 88
References / 89

5 The Operator-Machine Interface **91**

Summary / 91
5.1 Field Device OMIs: Built-In / 92
5.2 Field Device OMIs: Portable / 93
5.3 Workstation OMIs: Hardware / 94
5.4 Workstation OMIs: Software / 96
5.5 Multivendor OMIs / 101
5.6 Future OMI Possibilities / 103
Bibliography / 104

6 Communication Systems **107**

Summary / 107
6.1 Communication Requirements / 109
6.1.1 The OSI Layered Architecture: A Mail Analogy / 110
6.1.1.1 Application Layer / 110
6.1.1.2 Presentation Layer / 111
6.1.1.3 Session Layer / 111
6.1.1.4 Transport Layer / 111
6.1.1.5 Network Layer / 111
6.1.1.6 Data Link Layer / 112
6.1.1.7 Physical Layer / 112
6.2 The Physical Layer / 112
6.2.1 Baseband Signaling / 113
6.2.2 Modulation Techniques / 115

6.2.3 Physical Media / 117
6.2.3.1 Common Cable Characteristics / 118
6.2.3.2 Twisted Pairs / 119
6.2.3.3 Coaxial Cable / 120
6.2.3.4 Fiber Optic Cable / 121
6.2.3.5 The Powerline / 124
6.2.3.6 Physical Layer Standards: EIA-232 / 125
6.2.3.7 Physical Layer Standards: EIA-485 / 127
6.3 The Data Link Layer / 128
6.3.1 Master/Slave Protocols / 129
6.3.2 Peer-to-Peer Protocols: Contention / 129
6.3.3 Peer-to-Peer Protocols: Token-Passing / 130
6.3.4 Network Topologies / 131
6.3.5 LAN Standards: IEEE 802.X / 133
6.3.6 LAN Standards: ARCNET / 136
6.3.7 LAN Standards: Ethernet/802.3 / 139
6.3.8 Data Link Error Detection / 144
6.3.9 Extending LANs: Repeaters and Bridges / 146
6.4 The Network Layer / 146
6.4.1 Extending LANs: Routers / 147
6.5 The Transport, Session, and Presentation Layers / 148
6.6 The Application Layer / 149
6.7 The Future of Networking / 150
References / 151
Sources for Standards / 151

7 ASHRAE's BACnet **153**

Summary / 153
7.1 What Is BACnet? / 154
7.2 Modeling DDCs as Network-Visible Objects / 154
7.3 BACnet Services / 158
7.3.1 Object Access Services / 160
7.3.2 Alarm and Event Services / 163
7.3.3 File Access Services / 165
7.3.4 Virtual Terminal Services / 166
7.3.5 Remote Device Management Services / 167
7.4 BACnet Encoding Concepts / 168
7.5 The Lower Layers / 174
7.5.1 The Network Layer / 175
7.5.1.1 Use of BACnet with IP and IPX
7.5.2 BACnet LANs / 182
7.5.2.1 MS/TP / 182
7.5.2.2 Point-to-Point / 186

7.6 Extensibility Issues / 189
7.6.1 The Private Transfer Services / 189
7.6.2 Adding Proprietary Properties to a Standardized Object / 190
7.6.3 Adding Proprietary Objects to BACnet / 190
7.7 Conformance and Specification / 190
References / 193
Sources for Standards / 193

8 DDC System Design: A Case Study **195**

Summary / 195
8.1 System Diagram / 195
8.2 Sequence of Operation / 198
8.3 I/O Summary / 200
8.4 Mode Summary / 200
8.5 Flow Diagram / 202
8.6 Control Software / 205
8.7 Conclusion / 205
References / 205

9 DDC Implementation **207**

Summary / 207
9.1 DDC System Design: A Few More Thoughts / 207
9.2 Specifying DDC / 208
9.3 Installing DDC / 214
9.4 Commissioning DDC / 216
9.5 Training / 218
9.6 Operation and Maintenance / 219
References / 222

Epilogue: The Future of DDC **223**

Glossary of DDC Abbreviations, Acronyms, and Terms **225**

Manufacturers of DDC Equipment **235**

Index **237**

Direct Digital Control of Building Systems

Chapter 1

ELECTRONICS
PCS
DATA COMMUNICATIONS
CONTROL THEORY
COMPUTERS
SOLID STATE PHYSICS
SENSORS/ACTUATORS
SEMI-CONDUCTORS

Introduction

SUMMARY

Distributed processing is clearly the wave of the future in computerized heating, ventilating, and air conditioning (HVAC) control, primarily because it offers the promise of both improved reliability and accuracy of regulation.

But distributed control should be viewed in its proper context, as an evolutionary development, made possible by the great leaps in control theory and computer technology that have characterized the latter part of the twentieth century—not as some startling innovation.

And the distribution of computing "intelligence" is not without its liabilities. Although application-specific, limited-function controllers are relatively "simple," they are still computers, with all the rights, privileges, and problems pertaining thereto. Other more centralized computer configurations, or *architectures,* have been used in the past and may provide an adequate solution to the control or monitoring problem to be solved.

To understand how far the building control industry has come, and where it is going, let's take a brief look at some of the historical developments that have ushered in the present era of distributed direct digital control (DDC) and some of the system architectures that have been used along the way.

1.1 HISTORICAL PERSPECTIVE

Not long ago I happened to see a remarkable documentary on public television. It dealt with the construction of the Panama Canal, that amazing collection of trenches, locks, lakes, and dams that effectively shortened the trade routes between east and west by thousands of miles.

The canal, it turns out, was *not* successfully completed on the first try; a previous attempt by a French team had failed miserably. What was the difference? What factors were present that permitted the latter group of engineers and workers to succeed where the previous group, by all accounts just as dedicated and well-financed, had failed?

The answer is that two critical pieces of the puzzle were missing in the first attempt. The first was the availability of railroad technology. Without a way to remove excavated soil from the digging area, the American canal builders would have doubtless suffered from the same problem that bedeviled their French predecessors: every time a tropical rainstorm lashed the work site, the recently excavated soil would turn to mud and wash back into the trench making the working conditions impossible and requiring the men to start over. The rail system not only allowed the soil to be removed, but the soil could be immediately transported to locations where earthen dams and dikes were being constructed to raise the level of Lake Gatun, the body of water which forms the central portion of the waterway.

The second critical factor was medical: the discovery that mosquitos were the carriers of malaria and yellow fever. The French effort had been mercilessly plagued by tropical illnesses that had decimated the work force, both physically and psychologically. The smart workers left; those who stayed often died. With morale at low ebb, the digging inexorably ground to a halt. The Americans, on the other hand, armed with their knowledge of the role of the mosquito, built housing with screens on the windows, eliminated standing water in the vicinity of the workers' residences, and took other measures to reduce the threat.

What on Earth does this have to do with DDC? Simply that DDC, like virtually all other major technological advances, would be impossible without the existence of just exactly the right combination of knowledge and technologies, of theory and hardware. In this respect DDC is like the Panama Canal, the Apollo missions to the moon, the Salk vaccine, and similar accomplishments: each is the result of weaving together strands of knowledge and specific technical capabilities to form the finished tapestry. The absence of any one strand, regardless of the enthusiasm and dedication of the weaver, would render the tapestry impossible!

On what areas of theory and practice does the existence of DDC depend? Here are some of them: automatic feedback control theory, electronics, mathematics, solid-state physics—particularly semiconductors, sensors, actuators, data communications, and, of course, computer technology in all its facets. Obviously not all of these areas are independent of each other and that is exactly the point. Major advances in technology usually involve the interdependence of many disciplines and advances in one area often spur advances in other areas. DDC applied to building control is a perfect example!

1.2 AUTOMATIC FEEDBACK AND EARLY BUILDING CONTROLS

Consider the evolution of automatic feedback control. DDC is simply an implementation of such control using devices and techniques that were not previously available.

Feedback control, the method of regulating a physical process by comparing the value of the physical quantity you wish to control with the desired value and then making adjustments to the process to achieve the desired state, has been around since the beginning of time. The sensors, actuators, and controllers were human and the process was undoubtedly heating: if a cave dweller needed to raise the temperature in the cave, the solution was to put more logs on the fire. This technique, of regulating the fuel supply, persisted with only minor refinements for countless millenia.

The first known instances of *automatic* feedback control date back to the third century B.C. where the Greek mechanicians Ktesibios, Philon, and Heron sought to control the level and flow of liquids. They developed—and documented—various types of float regulators. The applications included water clocks, oil lamps, and, would you believe, automatic wine dispensers (Mayr 1970).

Leonardo da Vinci was responsible for designing a number of mechanical regulators in the fifteenth century that found use in controlling the speed of mills powered both by wind and water.

As the science of metallurgy advanced, it became possible to contain fire in metal furnaces. In 1620, Cornelis Drebbel, a Dutchman, designed a thermostatic damper control which relied on the expansion and contraction of a metal spring to open and close the air damper door in response to temperature changes.

But the invention that probably played the most significant role in setting the stage for the industrial revolution was the steam engine. The advent of metal furnaces coupled with the design of a metallic vessel in which water could be converted to steam led to the possibility of an engine that could, in turn, convert chemical energy to rotary motion. This allowed machines to be developed that could be located anywhere, not just near sources of hydropower or muscle power! This was an invention destined to revolutionize both transportation and industry. But how to control the engine's speed? James Watt's answer, in 1769, was the flyball centrifugal governor, a device that used the balance between gravity and centrifugal force on a pair of weights to control the amount of steam allowed to enter the engine's pistons.

As the industrial revolution "picked up steam," James Faraday was probing the new science of electricity. His work on magnetic induction led to the development of a practical electric motor and in 1854 the first U.S. patent for a motor-driven rotary fan was awarded (Skolfield 1957).

But electricity was not widely available except from low-power, liquid batteries. It took Edison's invention of the incandescent light bulb in 1879 to spur the concept of centralized electricity generation and widespread distribution. As cities began to grow and architects and engineers perfected the multistory building, so too did the need for distributed heating and ventilating. Up until the early 1800s, building heating was strictly by means of radiant stoves. Forced-air heating (called "hot blast" heating in those days) was experimented with using fans powered by steam engines, hydropower or, much later, electricity. The desirable side effect was ventilation (Carpenter 1906).

In the latter part of the nineteenth century, steam and hot water systems became more common. Early versions of these systems made no provision at all for automatic control of temperature in the living spaces. The only control was the manually operated radiator supply valve—and the window! (We still have a few old buildings at Cornell, built in the late 1800s, whose only advancement from the technology just described has been the installation of automatic control valves on the radiators!)

The revolution in the control of building heating systems really began at the beginning of the twentieth century with the perfection of the thermostat. The entire temperature "regulator" consisted of three components: a thermostat, some of whose

> parts will move because of change of temperature in the surrounding air, the motion so produced being used either directly or indirectly to open dampers or valves, and thus to control the supply of heat. Second, means of transmitting and often multiplying the slight motion of the parts of the thermostat produced by change of temperature in the room, to the valves or dampers controlling the supply of heat. Third, a motor or mechanism for opening the valves or dampers, which may or may not be independent from the thermostat. (Carpenter 1906, p. 416)

Actuator motors, when they were present, were spring-wound clock-type drives, or driven by water, electricity, weights, or compressed air. Thermostats used to turn such motors on or off were called *positive* while those used to modulate the position of an actuator were called *gradual.*

One of the first pneumatic air compressors was developed by the Johnson Temperature Regulating Company in the late 1800s. It had a bellows mechanism that was powered by water! The air pressure so developed was then fed to thermostats that controlled valves and dampers in a way that has changed little in the last hundred years!

1.3 MANUAL CENTRAL CONTROL SYSTEMS

A significant advance in building controls occurred in the early 1960s when Honeywell announced the development of a line of pneumatic sensors capable of measuring temperature and pressure and converting the result into an air pressure proportional to the measured value. Other vendors quickly followed suit, and the stage was set for the introduction of control panels where measurements from an entire air handler could be displayed at a single location. Where earlier instruments for tem-

ability to go the other way, from a number to a voltage or current, was also needed. The third was to reach agreement on what the various zeros and ones should mean in terms of what was being asked for or sent back in response.

The first problem was solved by the development of *multiplexers,* circuitry that allows an individual sensor or actuator to be accessed independently of all the others.

Circuitry carrying out what is called *analog-to-digital* and *digital-to-analog* conversion provided the solution to the second problem.

The solution to the third problem was the development of conventions on how to compose messages with various meanings. Such conventions on the structure of computerized messages form the foundation of data communication *protocols.*

The effect of these three developments in electronics and data communications was to spur the introduction of field panels with as many designations as the manufacturers could devise. Some of the names were remote terminal units (RTUs), data gathering panels (DGPs), field interface devices (FIDs), or, simply, multiplexers (MUXes).

Most of these direct ancestors of today's DDC panels are now simply referred to as "dumb MUXes." The reason is that they all shared one common limitation: none of them contained a computer. Thus they could do no more than slavishly respond to the commands of a (presumably) more intelligent master computer that would tell them what sensor value to send back or which output device to actuate. Control systems built around such centralized intelligence could perform great deeds—until the central computer died. . .

1.6 STAND-ALONE DDC

It is hard to think of a device that has played a greater role in shaping the latter part of the twentieth century than the microprocessor. The culmination of years of work in solid-state physics and semiconductor fabrication technology, the microprocessor can be found nearly everywhere—in everyday devices from autos to wristwatches. It is not surprising, therefore, that the controls companies saw its potential for application to building control!

In fact, the idea of embedding an "intelligent" microcomputer in a "dumb" MUX was so natural that virtually all the vendors jumped at the chance. Fortunately, there were precedents for them to follow since the industrial process control manufacturers such as Bailey, Fisher, GE, Taylor, and Westinghouse had already blazed the trail.

There were, nonetheless, significant problems, both technical and nontechnical. The technical ones concerned developing the inhouse expertise in electronics and computers to actually design and build viable products at a price that building owners could afford. The major nontechnical problem centered on the reeducation of personnel: entire corporate infrastructures needed to be retrained and reoriented. Plastic tubing was to give way to wire, pressure gauges to voltmeters, and piping diagrams to software printouts. As in any revolution, for some it was too much. But most persevered and in time DDC came to be accepted, not only simply as a viable alternative to time-honored pneumatic controls, but as the control technology of choice for reasons of power, flexibility, and economy.

1.7 DISTRIBUTED DDC

But the story of DDC does not end with the marriage of microcomputers and field panel multiplexers; there is one more wedding yet to describe. It is the marriage of DDC and data communications, specifically local area network (LAN) technology.

Although stand-alone processors completed the challenge of providing a microcomputer replacement for the previous pneumatic receiver-controllers, thus being particularly well suited for use in small buildings or other applications of limited complexity, there was still a need to interconnect computers on a building- or facility-wide basis. This desire for *interoperability,* the ability to share data among different computers, perhaps even computers from different vendors, has been one of the driving forces of the "information age."

In the case of building controls, the ability to network computers allows systems to be designed that represent a truly integrated solution to the problem being solved: sensor values can be shared, time schedules can be synchronized, global energy optimization strategies can be implemented, and multiple operator workstations can be connected. All of these measures, and more, lead to the possibility of more rational and efficient building system operation.

Note the use of the word *possibility.* Unfortunately, LANs, by themselves, do not provide the entire solution to the interoperability problem, a problem exacerbated by the advent of even smaller *unitary* or *application specific* controllers designed to further distribute intelligence out to the air distribution box or zone thermostat level.

What LAN technology *does* provide is the ability to get messages from one computer to another, much as the telephone allows you to call virtually anyone in the world who is similarly equipped. The problem is that *communication* can only occur if both parties speak the same language. In the computer world, the language is defined by a communication protocol which sets forth the vocabulary, syntax, and grammar of the messages which are to be exchanged and mutually understood.

Just as in the world of people, distributed DDC systems have historically used a multiplicity of tongues, a different one for each vendor's equipment. And again as in the world of people, this language barrier has limited commerce, communication, and building owners' confidence in the security of their DDC investment. Fortunately, there is now realistic hope that the situation is about to change, thanks to work being carried out under the auspices of the American Society of Heating, Refrigerating and Air-Conditioning Engineers (ASHRAE) which seeks to establish a *standard* communication protocol. This new protocol is called BACnet (for *B*uilding *A*utomation and *C*ontrol *Net*work). If it is successful, it will open up a whole new world of distributed building control possibilities.

In the chapters that follow we will look at each of the constituent technologies that go into making distributed DDC a reality.

REFERENCES

Carpenter, R. C., *Heating and Ventilating Buildings*, New York: Wiley, 1906.

Mayr, O., *The Origins of Feedback Control*, Cambridge, MA: MIT Press, 1970.

Skolfield, W. K., *A Century of Electric Fans*, Bridgeport, CT: General Electric Company, 1957.

Chapter 2

Sensors and Actuators

SUMMARY

Sensors and actuators are the eyes, ears, and hands of a computer control system. Sensors consist of an element or structure whose electrical or physical properties change in some direct and predictable way with variations in the quantity being measured. Since these property changes may be small and nonlinear, circuitry is often required to detect, amplify, linearize, and convert these signals into standard voltages or currents that can be easily sampled and manipulated. This process is known as *signal conditioning* and may be carried out in the vicinity of the sensor or within a field panel where several sensors may share common electronics.

Some quantities are measured indirectly. Airflow, for example, can be inferred by sensing its cooling effect on the temperature of a wire heated by a known amount of current or by carrying out a calculation based on pressure measurements.

Sensor/signal-conditioner pairs are chosen based upon the stringency of an application's requirements which, ultimately, strongly influence the cost. (You pretty much get what you pay for!) Factors to be considered include range, sensitivity, linearity, response time, accuracy, repeatability, interchangeability, ease of calibration, and stability, among others.

Signal-conditioner/actuator pairs carry out the inverse function in that an electrical signal is transduced or converted into a particular physical output quantity which is then used to physically move a valve, damper, or relay. The same selection factors previously cited also apply.

2.1 SENSORS

Sensors provide the means to sample the condition of physically measurable quantities in the world around us in such a way that the information can be processed by machines. In the case of sensors used in DDC, this usually means that the sensor possesses some characteristic which converts, or *transduces* the state of the measured variable into some electrically measurable quantity such as resistance, capacitance, inductance, voltage, or current.

There is, incidentally, considerable confusion in the literature on the definitions of the terms *sensor* and *transducer.* Since all definitions agree that both convert some physically measurable quantity into some other physically measurable quantity, the one thing that can be stated with assurance is that a sensor is the device in direct contact with the process whose properties we are trying to measure. If several stages of transduction are required to convert the output of a sensor into an electrically measurable quantity, the device is more likely to be called a transducer.

An example would be a pressure sensing device consisting of a bellows connected to a strain gauge whose resistance changes in response to the deformation resulting from the force exerted on it by the bellows. In this case, the bellows produces a force signal which is transduced by the strain gauge into a resistance signal. The key point is that a sensor and a process interact in such a way that an electrical signal directly related to the process can ultimately be developed, possibly after several stages of transduction from one form of signal to another!

Often, the output of a transducer is still not in a form suitable for a computer. It may need to be filtered or averaged over time, amplified, linearized, or converted into a standard signal range such as 4 to 20 milliamps (mA). Moreover, although sensors themselves are usually analog devices, with properties that change in a continuous way in response to changes in the measured variable, there may be a need to produce a binary output, such as a contact closure, from the sensing process. All these functions are aspects of signal conditioning which we discuss in Section 2.2.

Since virtually anything can be sensed to nearly any degree of accuracy, the problem usually comes down to selecting the most appropriate device from the competing possibilities. Which sensor is "best" depends very much on the application at hand. Here are some factors that should be considered:

1. **Range:** Over what range of the input variable is the sensor designed to operate? Is the sensor equally accurate over its entire range?
2. **Sensitivity:** How much will the input variable have to change for the sensor to produce a measurable change in its output?
3. **Linearity:** Will the sensor output change linearly with changes in the measured variable? If not, what type of signal-conditioning equipment or calculation is required to derive a linear result?
4. **Response Time:** How rapidly will the sensor output change for a given change in the measured variable? Is this fast enough for the application?

5. **Accuracy:** How closely will the sensor output relate to the actual value of the measured quantity?
6. **Repeatability:** Will the sensor output be repeatable for the same value of the input variable?
7. **Interchangeability:** Can the sensor be replaced without the necessity of having to recalibrate any signal-conditioning equipment?
8. **Ease of Calibration:** Will it be easy to establish and maintain the calibration of the sensor?
9. **Long-Term Stability:** Will the sensor output be likely to drift over time? How often will recalibration be necessary to attain the desired performance?
10. **Cost:** Is the cost appropriate for the value of information gained from the sensor?

In HVAC&R applications, we are generally trying to measure the characteristics of air, water, steam, and electricity, the most relevant of which are temperature, pressure, velocity, and flow. The sensors for each type of measurement will be described next.

2.1.1 Temperature Sensors

Temperature is the most commonly measured quantity in HVAC&R applications and its measurement probably has the longest history, dating back at least as far as Galileo. He is credited with developing the first liquid tube thermometer around 1592 (Benedict 1984). The four most common *electrical* temperature sensors are of far more recent invention. They are resistance temperature detectors, thermistors, thermocouples, and solid-state integrated circuit (IC) sensors.

The *temperature* of matter, whether solid, liquid, or gas, is a measure of its heat energy content, i.e., its degree of hotness or coldness. At the atomic level, heat energy translates into the movement of molecules. The *absolute temperature scale,* developed in the early 1800s by Lord Kelvin, is based on the physical properties of an ideal, frictionless gas. *Absolute zero* is the absolute temperature at which all molecular motion ceases. Other temperatures are measured using this scale in degrees Kelvin (K) where 1 K is $\frac{1}{100}$ of the temperature span between the freezing and boiling points of water. The other two common temperature scales are, of course, Fahrenheit (°F) and Celsius (°C). The relationships between the scales are

$$\begin{aligned} °C &= K - 273.15 \\ °F &= \tfrac{9}{5}°C + 32 \end{aligned} \tag{2.1}$$

The worldwide calibration of temperature sensors is based upon the temperatures of a number of "standard" physical phenomena. These reference temperatures are specified in a National Institute of Standards and Technology document called the International Practical Temperature Scale. Last revised in 1968, the document specifies 11 temperatures ranging from the triple point of hydrogen (13.81 K) to the freezing point of gold (1337.58 K).

2.1.1.1 *Resistance Temperature Detectors* Resistance temperature detectors (RTDs) are precision metallic resistors whose resistance changes in a known and repeatable way with changes in temperature. The resistance of an RTD as a function of temperature can be expressed as a polynomial:

$$R = R_0(1 + \alpha_1 T + \alpha_2 T^2 + \cdots) \tag{2.2}$$

where R is the resistance at temperature T, R_0 is called the *nominal resistance* and the α's are called the *temperature coefficients of resistance.* For RTDs R_0 is traditionally specified at 0°C and is the value used to reference a particular RTD made of a given metal, e.g., 100-Ω platinum. For most RTD metals, $\alpha_2 \approx 10^{-3} \times \alpha_1$ so that for most practical purposes Eq. (2.2) reduces to

$$R = R_0(1 + \alpha T) \tag{2.3}$$

where α can be thought of as the slope of the R vs. T curve divided by the nominal resistance R_0. In the linear approximation given by Eq. (2.3), α is assumed to be a constant and is traditionally evaluated over the range of 0 to 100°C. In this case:

$$\alpha = \frac{(R_{100} - R_0)/100}{R_0} \tag{2.4}$$

RTDs are made from both pure elements and alloys. The pure elements most commonly used are nickel and platinum while the most common alloys are mixtures of nickel and iron, a prime example being BALCO, an alloy of 70% nickel and 30% iron manufactured by Carpenter Technology Corporation. Platinum is probably the most preferred pure element because it is relatively inert chemically and does not oxidize in the presence of air. This allows it to be more easily purified than metals with a higher tendency to oxidize. It is the presence of impurities, and the compounds formed between them and the base metal of the RTD, that makes it difficult to guarantee the resistivity of RTDs made of other materials from one manufacturing run to the next to the same degree as is possible with platinum.

Two fabrication methods are widespread. The first involves winding a length of wire of the chosen material and wire gauge onto a thermally insulating, often ceramic, bobbin. The second consists of depositing a thin metallic film onto an insulating substrate, then using a laser to carve out a pattern of the desired length and shape. Calibration is accomplished by altering the number of turns on the bobbin or adjusting the pattern of the deposited film. See Fig. 2.1. Lead wires are then attached, and the calibrated assembly is sealed or coated with epoxy, glass, or some similar material to protect it against mechanical or chemical damage. The finished RTD can then be installed in a probe package appropriate to the final application.

The new thin-film RTDs have several advantages over the older wire-wound type. The first is that they are incredibly small, measuring only a millimeter or two on a side. This means that they can respond more rapidly to temperature changes compared to the larger, more massive wire-wound devices. Second, thin-film RTDs can

FIGURE 2.1. A photomicrograph of a 1000-Ω platinum RTD. Notice in the lower right how the circuit was laser-etched to increase the RTD's overall resistance and bring it to specification. (Courtesy HY-CAL Engineering.)

be fabricated with a higher nominal resistance than wire-wound RTDs of comparable size. The higher nominal resistance can be important where long lead wires are involved, as we shall see shortly. The smaller size is significant primarily in non-HVAC applications. Finally, thin-film RTDs can be made for less money, in part because less metal is used.

A disadvantage of the small size of thin-film RTDs is self-heating. This occurs because the only way to determine the resistance of an RTD is to pass a current through it. The resultant voltage drop results in an ohmic power dissipation of I^2R which causes the RTD's temperature, and resistance, to rise. The significance of self-heating depends on the ability of the medium with which the RTD is in contact to dissipate the resultant heat. The effect can be minimized by using heat sinks, conductive grease, and the smallest possible sensing current.

The main challenges in manufacturing RTDs stem from the metallurgy involved and the physical mounting of the device. The metallurgical challenge is to be able to produce the metallic sensor elements with precisely the same composition from batch to batch. If the composition or impurity content vary, the resistivity of the metal will change, making it more difficult to produce truly interchangeable sensors. The physical mounting is important because it affects the amount of deformation the sensor experiences as it expands or contracts with temperature changes. These deformations can change the cross-sectional area of the RTD conductors resulting in nonlinear resistance changes.

The largest potential source of error in using RTDs stems from a failure to consider the resistance of the RTD lead wires. To see the effect of lead wire resistance, let's look at the basic circuit used in resistance measuring, the Wheatstone bridge. Figure 2.2 shows the classical circuit set up for manual operation. Suppose that $R_1 = R_2$. In this case, the voltage drop across each resistor on the left side of the bridge is one half the supply voltage, $V_i/2$. The bridge is said to be balanced when the same drop is present across R_x. At this point, $R_v = R_x$ and $V_o = 0$. Operated manually, the variable

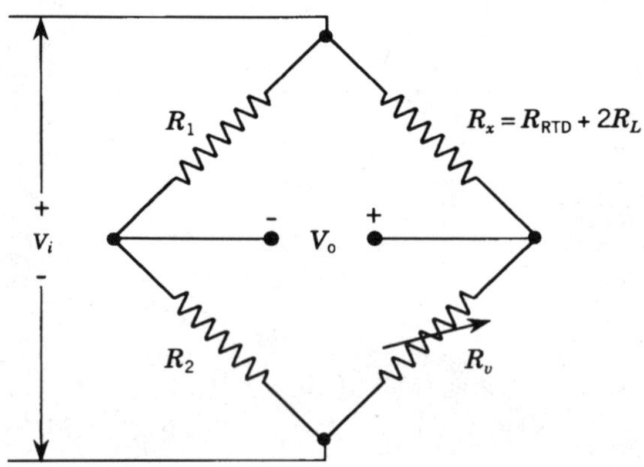

FIGURE 2.2. The basic Wheatstone bridge circuit used to determine the value of an unknown resistance. If $R_1 = R_2$, then $V_o = 0$ when $R_v = R_x$.

resistor R_v would be adjusted to obtain a null on the voltmeter measuring V_o. The resistance of R_x could then be read off the dial of R_v.

The more appropriate situation to discuss occurs when the bridge is *not* balanced, as is the case in practical signal-conditioning circuits. See Fig. 2.3. Here all the bridge resistors have the same value except for R_x. Applying Ohm's law to each side of the bridge we can derive an expression for R_x as a function of R, V_o, and V_i:

$$R_x = \frac{V_i - 2V_o}{V_i + 2V_o} R \tag{2.5}$$

This equation gives us a way of directly relating the unknown resistance to a measurable voltage, which can then be processed and fed to a computer.

Note that in either case, what we end up with is not equal to R_{RTD} but rather $R_{\text{RTD}} + 2R_L$ where $2R_L$ is the resistance of the two lead wires between the RTD and the bridge.

How significant is this error in practice? Consider a platinum 100-Ω RTD. The most common α for this type of RTD (corresponding to the German Industry Standard DIN 43760) is 0.00385. This means that a resistance change of 0.385 Ω represents approximately a 1°C change in temperature. Suppose our lead wires are 20 gauge copper. How long would the lead wires have to be to throw off our reading by 1°C? Since this type of wire has a resistance of about 0.0103 Ω/ft (at 25°C), it would only take a distance of about 19 ft from RTD to bridge to produce this error (0.385/0.0103/2)!

One way to improve the situation is to use an RTD with a higher nominal resistance. Suppose we chose a 1000-Ω platinum sensor. The α is the same as with the 100-Ω sensor, but, from Eq. (2.4), the average resistance change per degree is now

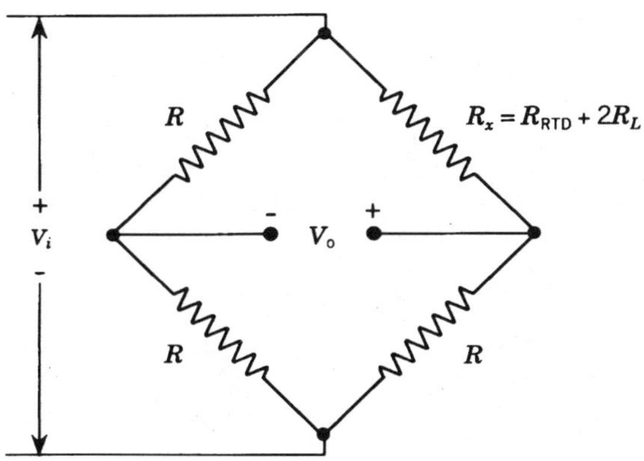

FIGURE 2.5. The Wheatstone bridge circuit of Fig. 2.2 can be modified for automatic use. Eq. (2.5) shows how R_x can be calculated from knowledge of V_o, V_i, and R. In practice, the bridge is part of signal-conditioning circuitry which eventually produces an output suitable for input into the DDC computer.

10 times higher, or 3.85 Ω/°C. In this case our same lead wires would only contribute an error of 0.1°C to our overall measurement.

The moral of the story is that it is a good idea to select RTDs with higher nominal resistances rather than lower, unless of course you can put the RTD and signal-conditioning equipment at the same place.

If, however, the RTD and signal-conditioner *cannot* be co-located, there are still at least two more ways of dealing with the effects of lead resistance. One is to use a so-called three- or four-wire RTD. A three-wire hook-up is shown in Fig. 2.4. The result of this arrangement is that two lead wires are now effectively in adjacent sides of the bridge while the third appears in the measurement circuit. If the bridge is balanced, this provides a perfect solution since the two R_L's exactly cancel and no current flows through the measurement circuit. If the bridge is *not* balanced, there will still be some error but approximately only half of what it was with the two-wire RTD. Further improvement is possible with four-wire RTDs and appropriate bridging, but such sensors are rarely used outside of laboratory settings.

The final, and certainly easiest, way of dealing with the problem is to adjust the database constants in the DDC that convert the "raw" voltage or current from the bridge/signal-conditioner into the final temperature reading. If you know you have the equivalent of 1° of lead resistance, you can simply have the computer subtract it from the RTD reading before it is used! But be warned: this technique should never be used to compensate for sensors that are out of tolerance, since there also may be other problems with such sensors. Also, such compensation is strictly valid at only one particular temperature, since lead wires themselves change resistance with temperature.

2.1.1.2 Thermistors Thermistors, by definition, are *therm*al res*istors*. Like RTDs, they are devices that exhibit a predictable, repeatable change in resistance when subjected to a change in their ambient temperature. Here the similarity ends.

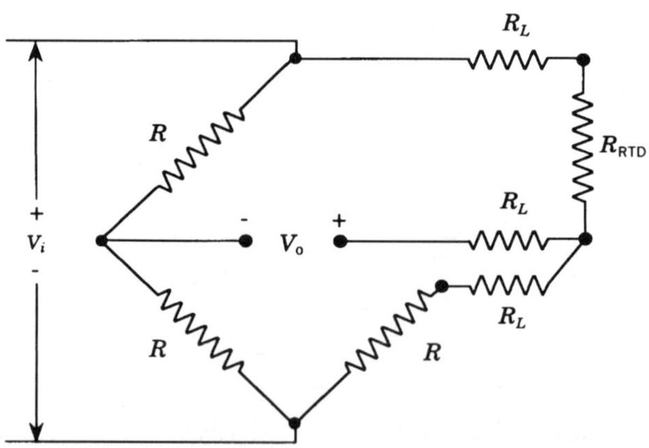

FIGURE 2.4. The Wheatstone bridge circuit resulting from the use of a three-wire RTD. At balance, $R_{RTD} + R_L = R + R_L$ but at other times the compensation is imperfect.

Where RTDs are composed of metals, thermistors are made of ceramic, semiconducting mixtures of oxides of metals such as manganese, nickel, cobalt, copper, iron, and titanium. Where the resistance of RTDs increases with an increase in temperature, the resistance of thermistors decreases with an increase in temperature, i.e., they have a *negative* temperature coefficient.

But the greatest difference between RTDs and thermistors, and the characteristic that makes them the device of choice for applications where high resolution is important, is their sensitivity—their change in resistance for a given change in temperature. Where an RTD may change a fraction of an ohm/°C, a thermistor may change 100 Ω! (An example of such an application is the BTU metering of chilled water where tenths of degrees in the supply and return temperature measurements can make a big difference in the resulting energy reading.)

Thermistors are made by preparing a mix of oxides with the desired metallurgical and electrical properties, forming the mix into the desired geometry, and sintering it. (*Sintering* is the process of fusing a powdered material by heating it, usually under pressure, until it solidifies.) As you might imagine, it is extremely difficult to predict the exact characteristics of the resulting device ahead of time since they depend on the precise composition, sintering time and temperature, and geometry, among other things.

Stability can also be a problem. Since thermistors are forged at high temperature, their characteristics can change if they are again subjected to an elevated temperature later on. Fortunately, both these factors are more of a problem for the manufacturer than for the user since thermistors can be obtained with resistance tolerances and stability specifications appropriate to nearly any application.

Thermistors are classified according to the method by which the lead wires are attached to their ceramic body. In the first group of devices, the lead wires are embedded in the body during the sintering process while in the second group the leads are attached to metallized surface contacts which are applied to the thermistor following its fabrication.

The thermistors in the first group are called "beads" while those in the second group are called "chips" or "flakes." Both types of device come with and without various protective coatings including glass and organic polymers. Bead-type thermistors can be used up to 550°C without danger of the leads falling off, whereas the chip-type devices are generally rated for operation at or below 150°C. The main advantage of the chip-type thermistors is that they can be trimmed after initial fabrication to "fine-tune" their characteristics.

The relationship between the resistance and temperature of a thermistor can be expressed by the equation

$$R = R_0 \, e^{\beta(1/T - 1/T_0)} \tag{2.6}$$

where R is the resistance at temperature T, R_0 is the resistance at T_0, e is 2.718 . . . , the base of the natural logarithm system, and β is a characteristic of the thermistor composition. β and the temperatures are all in units of degrees Kelvin. Most thermistor specification sheets give values of R_0 at 25°C. See Fig. 2.5.

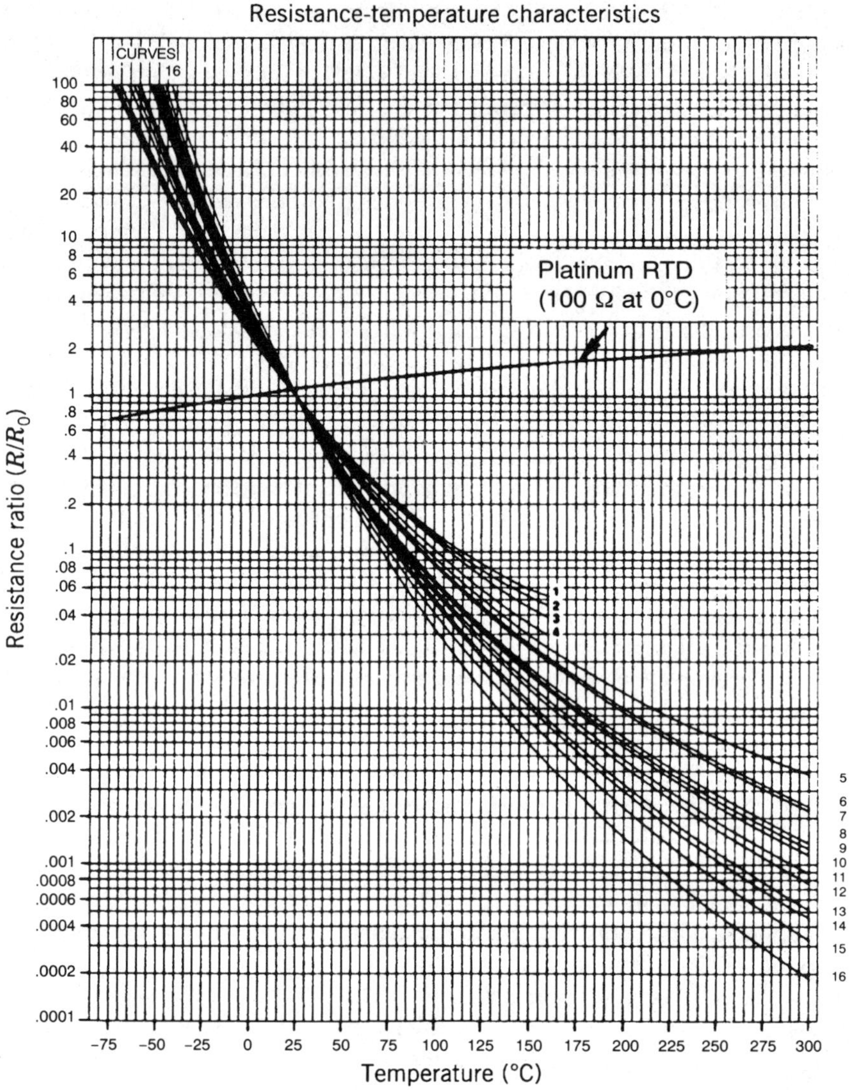

FIGURE 2.5. A comparison of the resistance/temperature characteristics of a typical RTD and family of thermistors. Each thermistor curve is for a different β. See Eq. (2.6). The curves are normalized with respect to each device's nominal resistance R_0. Note that for the RTD, R_0 is evaluated at 0°C while for the thermistors R_0 is the resistance at 25°C. At any temperature, a device's resistance is obtained by multiplying the resistance ratio by the device's nominal resistance. Thermistors are available with R_0's ranging from hundreds to millions of ohms.

The circuitry used to measure R is often similar to that used for RTDs. Because of thermistors' small size and the potentially large self-heating effect, attention is always given to using the minimum possible sensing current. This is driven home in the commercial literature where R_0 is usually referred to as the "zero power" nominal resistance.

2.1.1.3 Thermocouples Thermocouples are temperature sensors that make use of a physical phenomenon discovered by Thomas Seebeck in 1821. He found that if two wires composed of dissimilar metals were joined at both ends (thus forming at each end "couples"), and one junction was heated relative to the other, a continuous current would flow through the resulting circuit, a fact now known as the *Seebeck effect.* See Fig. 2.6. If the same circuit is broken in the middle, the resulting open circuit voltage in the direction of either junction is a function of the temperature of the junction and the compositions of the dissimilar metals. Thus, while RTDs and thermistors change *resistance* in response to temperature swings, thermocouples generate a *voltage* signal.

Since all dissimilar metals exhibit the Seebeck effect to varying degrees, the challenge over the years has been to find the combination of metals with the most desirable properties for the application at hand. One of the most obviously desirable properties for this type of device would be for it to generate the largest possible voltage change for a given temperature change. For small temperature spans, the relation between voltage and temperature can be approximated by a straight line such that

$$\Delta V_{AB} = \alpha \, \Delta T \tag{2.7}$$

where ΔV_{AB} is the voltage change between dissimilar metals A and B for a temperature change of ΔT and α, the constant of proportionality, is called the Seebeck coefficient. For the purpose of Eq. (2.7), α is evaluated somewhere in the range of the temperatures making up the ΔT where, if ΔT is sufficiently small, α may be considered constant. In the commercial literature α is usually specified at 0 or 600°C.

To give you some feel for the size of the effect we are considering, and the difficulty in designing circuitry to accurately measure it, typical values for the

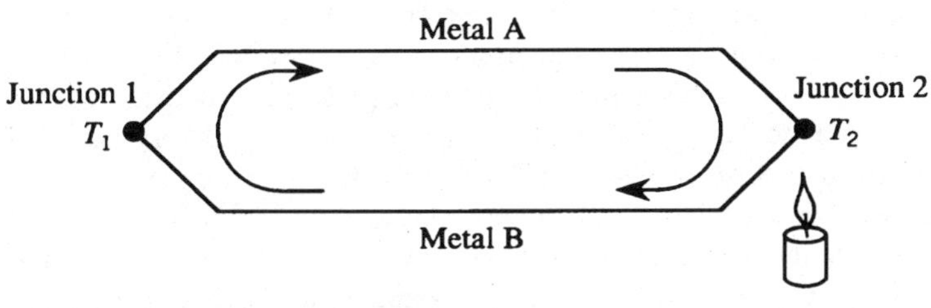

FIGURE 2.6. A Seebeck current is produced when dissimilar metals are joined to form two junctions at different temperatures.

Seebeck coefficient are in the range of 10 to 80 μV/°C (*micro*volts per degree Celsius)! For a thermocouple with an α of 50 μV/°C at 0°C, for example, this means that the total voltage change between 0 and 25°C would only amount to 0.00125 V, a mere 1.25 mV. On the positive side, thermocouples can be used over a huge temperature range, from subzero to several thousand degrees. And, consisting as they do of two wires bonded together at a single point, they are extremely rugged and, if need be, can even be fabricated in the field.

Since, technically speaking, the coupling of any two dissimilar metals produces a thermocouple, the need quickly arose to develop standards for the characteristics of these sensors. The "holy grail" of thermocouples is the American National Standard ANSI/MC96.1-1982 "Temperature Measurement Thermocouples," available from ANSI or the Instrument Society of America (ISA). The latest 1982 revision dates back to the work of the ANSI Committee on Temperature Measurement, MC96, which was organized in 1946 under ISA sponsorship.

The essence of the standard is the specification of seven temperature-voltage relationships and corresponding tables showing millivolt output versus degrees Celsius, each of which is identified by one of the letters B, E, J, K, R, S, or T. Thermocouple metal pairs capable of producing the specified temperature-voltage curves are also given although, in theory at least, other materials could also be used if they produced the same results. For each lettered type, the standard also contains nominal temperature ranges, calibration tolerances, and application recommendations.

The practical problem with thermocouples is that the dissimilar metals must, sooner or later, come into contact with other metals so the junction voltage can be measured. If the thermocouple wires and the terminals of the voltmeter (or other intermediate circuitry) are of different metals, these junctions too will produce a thermoelectric voltage that must be taken into account.

One way to solve the problem is to break the original thermoelectric circuit of Fig. 2.6 along the side and bring these wires to the voltmeter. Assuming the voltmeter terminals are at the same temperature, the voltage contribution from their contacts with the thermocouple wire will be equal and opposite. See Fig. 2.7. The difficulty is that, having solved one problem, we have introduced another. We now have *two* junctions in the circuit instead of just the one we are interested in and are actually measuring the difference in voltage between the two. This second problem is solved if we happen to know the exact temperature of the second junction since we can then subtract out its contribution to the measurement. In fact, the second junction is called the *reference junction* and most thermocouple tables actually show the voltage obtained with such a reference junction maintained at the ice bath temperature of 0°C.

Happily, in practice, the necessity for an actual ice bath can be eliminated by various hardware and software means. Hardware compensation, as it is called, can be designed using a secondary temperature sensor such as an RTD to measure the reference junction temperature and generate a voltage in the circuit equivalent to what a 0° junction would produce. The reference junction in this case may simply be a screw terminal connection on a terminal block. But why not use the RTD to measure what you want to measure in the first place? The answer is that the thermocouple may

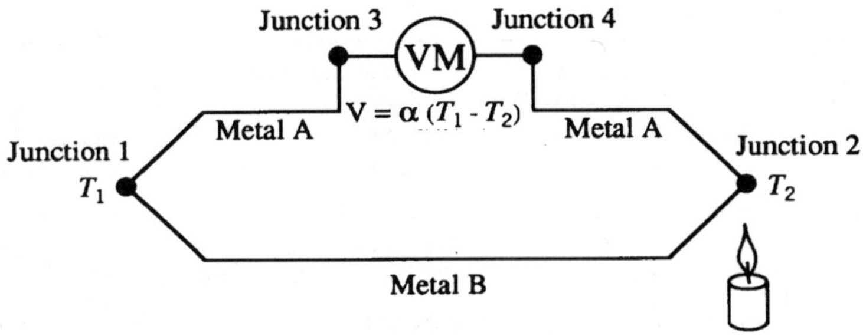

FIGURE 2.7. If one leg of the circuit is broken, the two additional junctions formed between the voltmeter leads, or other measuring circuitry, and metal A will have no effect as long as they are at the same temperature.

be at a temperature or in an environment that would wipe out the RTD. Also, the RTD (or thermistor, or integrated circuit sensor) only has to work around the temperature of the reference junction, which is likely to be in the vicinity of room temperature.

In software compensation, the reading of the secondary temperature sensor and a knowledge of the hardware interconnections are used to mathematically calculate the effect of the various junctions on the measured sensor voltage. As a result, the computer can mathematically infer what the thermocouple junction voltage, and hence temperature, must be.

2.1.1.4 Integrated Circuit Temperature Sensors A relatively recent development is the solid-state, integrated circuit temperature sensor. These devices are usually based on the voltage-current characteristic of reverse-biased semiconductor diode junctions. Such junctions display a linear voltage-current relationship as a function of temperature, although the amount of current is minuscule. Commercial sensors are available as either current or voltage source devices. A typical rating for a current source device is 1 μA/K while a voltage source device might be rated at 10 mV/K. The devices are usually calibrated such that the number of microamps or millivolts corresponds directly to the absolute temperature, e.g., a reading of 273.15 μA would be equal to 273.15 K or 0°C.

Although IC sensors are relatively inexpensive, they have some drawbacks. Among them are limited temperature range operation compared to other sensors (only –50 to 150°C), relative fragility, lack of stability over time, and they require an external power supply. Add to these limitations the fact that few manufacturers have chosen to package IC sensors in the types of probes most used in building system DDC applications. The bottom line is that IC sensors are not widely used in general-purpose HVAC&R applications.

2.1.1.5 Temperature Sensor Summary Selecting the appropriate temperature sensor, like most engineering, involves trade-offs. There is no perfect sensor for all DDC applications! See Table 2.1. My own choice would be to use RTDs with standard nominal resistances in the 1000 Ω vicinity for most measurements, choosing

TABLE 2.1 Temperature Sensors

Sensor Type	Primary Use	Advantages	Disadvantages
RTD	General purpose air, water, steam	Very accurate, interchangeable, stable	Relatively expensive, not very sensitive
Thermistor	High sensitivity applications, chilled water metering	Large change in resistance for small change in termperature	Nonlinear, fragile, self-heating can be a problem
Thermocouple	High-temperature applications, boilers, stack gas	Inexpensive, self-powered, rugged	Low-voltage output, not very sensitive
I. C. Sensor	General-purpose, low-temperature applications ($T < 200°C$)	Very linear, high output, inexpensive	Not very rugged, limited physical configurations available

platinum, with its higher cost, over nickel-iron for the most critical. But beware of "odd-ball" nominal resistances. Using such sensors could result in interchangeability problems later on or problems replacing the associated signal conditioners.

For low temperature metering applications, such as chilled water, I would choose the highest grade thermistors I could find because of their high sensitivity and ability to respond, repeatably, to small temperature changes. For high-temperature applications there is no substitute for thermocouples. Don't forget that the overall quality of your measurements depends not only on the sensors but also on any subsequent signal-conditioning equipment.

2.1.2 Pressure Sensors

After temperature, the most commonly measured quantity in HVAC&R systems is pressure. Pressure is important not only in and of itself. It is also commonly used to infer velocity and flow, a subject we cover in Section 2.1.3.

While the physical definition of pressure is simply force divided by area, the engineering units used to indicate specific pressure values are, for historical reasons, exceedingly numerous. Among these are pounds per square inch (psi), Pascals (Pa) (each of which is 1 Newton/m^2 or 10 dynes/cm^2), and units derived from the heights of columns of various fluids such as air, water, and mercury. Table 2.2 contains conversion factors for the most common units used in the United States, including atmospheres, inches of water column (usually written "in WC"), and inches of mercury.

Apart from the range of pressure to be measured, it is also necessary to specify the desired pressure reference. Three such references are in widespread use. The first is an absolute vacuum. Pressures so referenced are called "absolute" and units such as pounds per square inch absolute, psia, are used.

The second reference is atmospheric pressure. Pressures referenced to it are called

TABLE 2.2 Common Pressure Conversion Factors

psi = inches WC × (3.6127×10^{-2})	inches WC = psi × 27.68
psi = inches Hg × (0.49118)	inches Hg = psi × 2.036
psi = mm WC × (1.4223×10^{-3})	mm WC = psi × 703.1
psi = mm Hg × (1.9339×10^{-2})	mm Hg = psi × 51.71
psi = cm WC × (1.4223×10^{-2})	cm WC = psi × 70.3
psi = kg/cm^2 × 14.233	kg/cm^2 = psi × 0.0703
psi = bar × 14.503	Bar = psi × 0.0689
psi = mbar × (1.4503×10^{-2})	mbar = psi × 68.95
psi = Pa × (1.4503×10^{-4})	Pa = psi × 6895
psi = kPa × (1.4503×10^{-1})	KPa = psi × 6.895
psi = atmospheres × 14.7	atmospheres = psi × 0.06804

"gage" pressures and the letter "g" is appended to the engineering units, e.g., psig. (It is conventional to use the spelling gage rather that gauge when discussing pressure measurement but no one seems to know why. We should, however, probably be grateful.)

In the third case, the reference pressure may itself be variable. This is usually the case when we are interested in measuring a pressure difference, such as exists across a fan, pump, or filter element. For such "differential" measurements, the letter "D" is added to the engineering unit abbreviation.

Pressure sensors consist of a force summing element, typically a diaphragm or bellows, which is displaced by a predictable amount by the imposed force and whose motion is resisted by a spring or other elastic structure. The problem then becomes one of transducing the resulting displacement into a measurable electrical quantity such as a change in capacitance, inductance, resistance, or voltage.

2.1.2.1 Capacitive Transducers Capacitive pressure transducers operate by coupling the displacement of a diaphragm or bellows to a change in separation of the two plates of a capacitor, thus altering its capacitance. See Fig 2.8. One plate of the capacitor is fixed to the body of the transducer while the other moves. The movable plate may even be the diaphragm itself. By using the transducer's capacitance as part of an oscillator circuit, for example, the associated signal-conditioning circuitry can relate the change in capacitance to a change in oscillator frequency. The oscillator output can then be readily converted to a standard voltage or current output.

2.1.2.2 Inductive Transducers Converting changes in displacement to changes in the mutual inductance between two wire coils is the underlying principle of the *linear variable differential transformer* (LVDT). See Fig. 2.9. LVDTs consists of a primary coil and two secondary coils wound around a hollow tube into which a ferrite shaft is fitted. The primary coil is excited by an alternating current. As the shaft moves, it couples differing amounts of the primary coil's magnetic field to each of the secondary coils, thereby inducing different voltages. The secondaries are wound in opposite directions around the core so that when the shaft is centered the

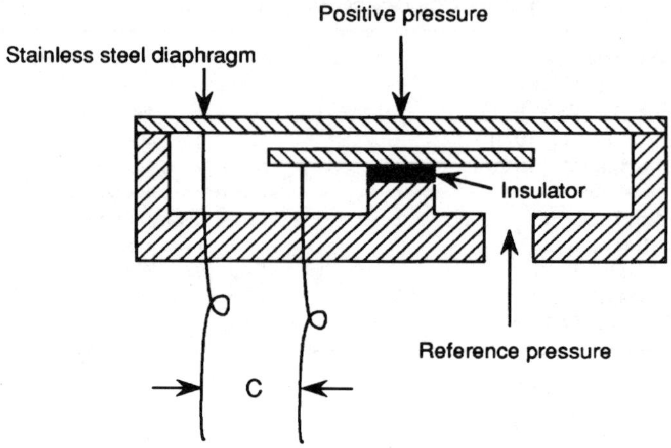

FIGURE 2.8. A capacitive pressure sensor.

voltages cancel out. The magnitude and phase of the resulting output voltage can thus be directly related to the position of the ferrite core and its related primary pressure-sensing element.

2.1.2.3 Strain Gauge Transducers A third common type of pressure transducer makes use of strain gauges. *Strain* is the deformation per unit length, $\Delta L/L$, that results when a body, such as a piece of wire, is subjected to a force or stress along its length. Since a change in length is accompanied by a change in diameter, the wire's electrical resistance also changes. The resistance change can then be measured by the types of bridge circuits discussed earlier which, as always, are built into the

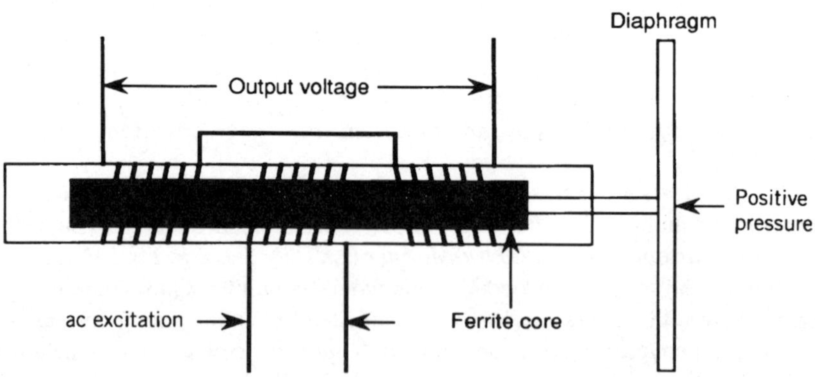

FIGURE 2.9. A linear variable differential transformer (LVDT) pressure transducer.

transducer's associated signal conditioning circuitry.

Two types of strain gauges are used in pressure sensing: bonded and unbonded. See Fig. 2.10. The difference between the two is the way in which the gauge interacts with the force summing element and is thereby subjected to stress. In the bonded case, the gauge is firmly attached to the element, which is deforming under the imposed pressure, and thus directly measures its deformation. In the unbonded case, the gauge is deformed by the displacement, rather than deformation, of the sensing element to which it is attached. Because of their ruggedness, bonded gauges are far more prevalent than the unbonded type. The most common of these are made by a printed circuit process using conductive alloys rolled to a thin foil. The foil is then etched to a grid configuration and attached to the pressure-sensing element with special adhesives.

2.1.2.4 Piezoelectric Transducers Another method of transducing pressure depends on the fact that the electrical characteristics of certain materials change when they are deformed under pressure. Such materials are called *piezoelectric* from the Greek word *piezein* meaning to squeeze. Crystalline quartz is a very common example. One such measurable characteristic is a change of resistance with pressure. Such substances are said to be piezoresistive, and the resistance signal they produce can be processed by the type of bridge circuits we have described earlier.

A more interesting effect results from the fact that certain crytalline or ceramic materials generate a small voltage difference across their structure when squeezed or, alternatively, deform slightly when subjected to a voltage difference. This behavior leads to the possibility of oscillation since energy can be shuttled back and forth between the imposed electric field and the potential energy stored in the deformed crystal. Since the vibration frequency of the crystal depends on both the frequency of the excitation voltage and the imposed pressure, it is possible, by maintaining a constant excitation frequency, to infer the pressure by measuring the crystal's frequency of oscillation.

2.1.2.5 Potentiometric Transducers A final technique uses a resistance potentiometer. A potentiometer is simply a three-terminal resistor with an adjustable center connection. A common type consists of a coil of wire wound on an insulating core. The center connection is in the form of a "wiper" which moves up and down the coil depending on the motion of its associated linkage. In a pressure transducer, the motion is determined by the force summing element. By measuring the resistance between the wiper and one end of the coil, the position of the wiper, and thus the displacement of the force summing element, can be determined.

2.1.2.6 Pressure Sensor Summary While all of the techniques just described are available in products, the devices most commonly used in HVAC&R DDC, particularly on the air side, are based on variable capacitance. The primary reason is cost! See Table 2.3. As always, check the requirements of the job and study the manufacturers' spec sheets to find the best, most cost-effective solution for your particular application.

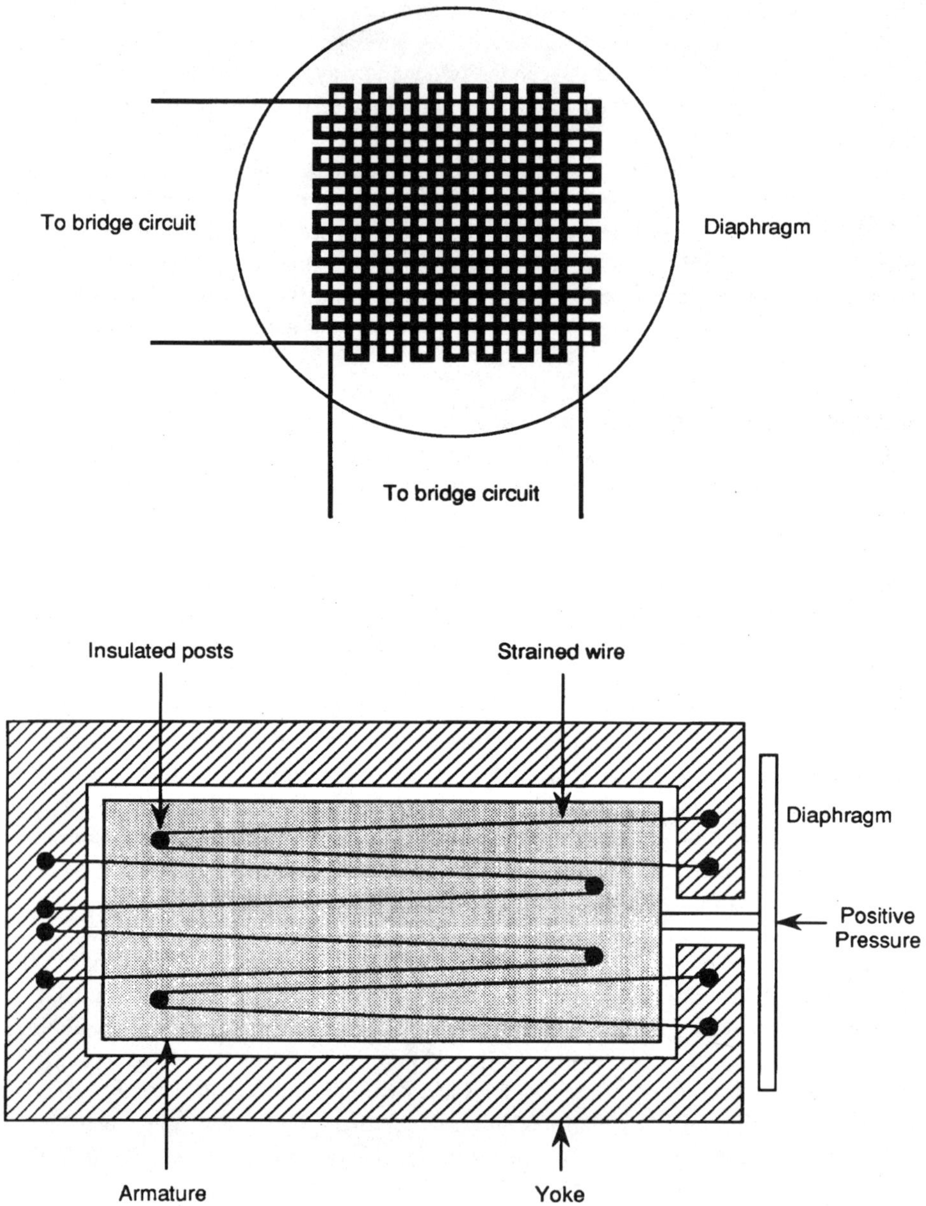

FIGURE 2.10. Bonded and unbonded strain gauges may also be used to measure pressure.

TABLE 2.3 Pressure Sensors

Sensor Type	Primary Use	Advantages	Disadvantages
Capacitive	Low-pressure air; duct static or filter ΔP	Inexpensive	Signal conditioning is complex, low output
Inductive	Low pressure, fume hood ΔP	Rugged construction	Expensive, temperature compensation may be difficult
Strain gauge	High pressure, chilled water, steam	Linear output	Low output signal
Piezoelectric	Fluctuating pressures, sound, mechanical vibration	Wide pressure range	May be difficult to calibrate using static load
Potentiometric	General purpose	Inexpensive, high output	Low accuracy, large size, physical wear may shorten life

2.1.3 Velocity and Flow Sensors

Since the control of most HVAC&R systems involves regulating the flow of air or water, it is not surprising that many techniques have been developed to make these important measurements. Happily, many velocity and flow measurements are derived from pressure measurements, a subject with which we are all now at least somewhat comfortable. See Table 2.4.

Velocity is a vector quantity whose magnitude is the time rate of change of position

TABLE 2.4 Velocity and Flow Sensors

Sensor Type	Primary Use	Advantages	Disadvantages
Pitot tube	Air	Inexpensive	Can plug up with dirt
Orifice plate	Water, steam	Inexpensive, available for many pipe sizes	Can erode, accuracy depends on diameter of orifice
Venturi tubes	Water, air	Lowest head loss of insertion devices	Relatively large size, more costly
Hot wire	Air	Measures mass flow, not easily contaminated	Somewhat fragile
Turbine	Steam	Good turndown ration	Blades can be destroyed by water slugs, moving parts wear
Vortex shedding	Water	Accurate	Complicated signal conditioning
Ultrasonic	Water	Nonintrusive	Most expensive

of an entity—speed—and which points in the direction of motion. It is measured in units such as inches/second, feet/minute, and meters/second. In HVAC&R work the directional aspect of velocity is usually implicit since the fluid whose velocity we are trying to measure or control is, hopefully, constrained to flow within the confines of a duct or pipe! *Flow* is the movement of a fluid expressed in terms of volume or mass. Common units in the United States are cubic feet per minute, gallons per hour, or, in the case of steam, pounds per hour.

To understand how fluid velocity is inferred from the measurement of pressure, a quick review of some basic physics is in order. The key is understanding two fundamental physical laws: conservation of energy and conservation of matter. (OK, it's true that matter can be converted to energy, $E = mc^2$ remember? But as long as we don't have any nuclear reactions going on inside our HVAC system we don't have to worry!)

Energy is the ability to do work. The law of conservation of energy states that in any closed system, energy can neither be created nor destroyed. The practical effect of this conservation law is that energy shuttles back and forth between various forms so that the *total energy* is always constant.

Although the equations describing the dynamics of fluids are generally complex, the conditions found in the air and water flows in HVAC systems allow us to greatly simplify the equations. We can do this by making three assumptions. The first is that the fluid is incompressible. This is a good assumption, even for air, as long as its velocity is low relative to the speed of sound (approximately 60,000 ft/min in dry air at standard sea level conditions). The second assumption is that the fluid is non-viscous. This also turns out to be reasonable for the fluids and conditions relevant to HVAC applications. The third is that the flow is steady over time. With these assumptions, the fluid equations reduce to a relation called Bernoulli's equation:

$$P + \rho\frac{V^2}{2} + \rho gh = \text{constant} \tag{2.8}$$

where P is the static pressure in the fluid at a given point, ρ is the fluid density (mass per unit volume), V is the fluid velocity, g is the gravitational constant, and h is the height of the fluid above an arbitrary datum. Bernoulli's equation is nothing more than a statement of the energy conservation law applied to liquids and gases. The static pressure term is a measure of the work done on the fluid per unit volume to pressurize it; the second term is the kinetic energy per unit volume and produces a "dynamic pressure" in the direction of the flow; and the third term is just the potential energy of the fluid due to its weight.

The second law needed to understand flow measurement is that of conservation of mass. It states, in effect, that what goes in one end of a pipe or duct must come out the other. If the conduit is of constant cross-sectional area, an ideal, incompressible fluid will flow at a constant speed throughout. But if the area varies, the speed of the fluid will also vary, slowing down where the area is bigger, speeding up where the area is smaller so that, at any moment, the same amount of fluid is passing through any given cross section. Mathematically,

$$\dot{m} = \rho A_i V_i = \text{constant} \tag{2.9}$$

where $\dot{m}$ is the mass flow rate, ρ is the fluid density, A_i is the cross-sectional area at location i, and V_i is the fluid velocity at location i. This equation, also known as the "continuity equation," says that if the area goes up the velocity must go down and vice versa. Armed with Eqs. (2.8) and (2.9) we can now look at how the most common flow sensors work.

2.1.3.1 Pitot Tubes Pitot tubes are used to measure air velocity in ducts (and past airplanes). The sensor/transmitter combination consists of a differential pressure transmitter and two tubes which may be fitted into a common housing. See Fig. 2.11. The pitot tube itself is inserted into the airstream and directed upwind. The second

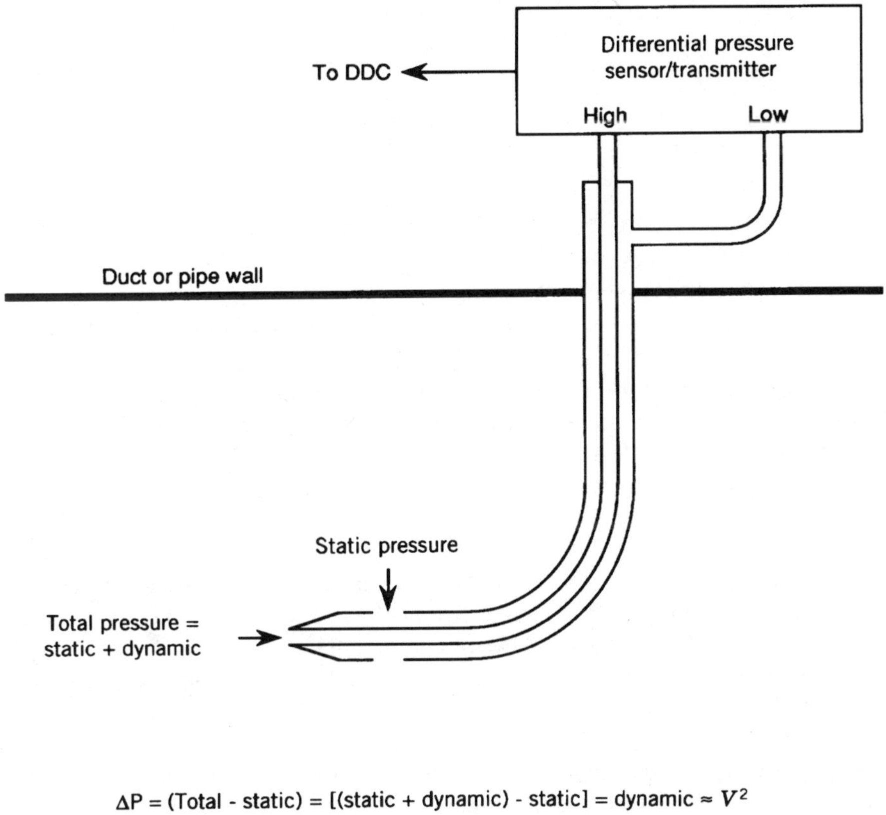

FIGURE 2.11. A typical pitot tube. In this case, the static and total pressures are sampled by the concentric tubes. Pitot tube arrays or slotted tubes that span the duct are also common but all make use of the same principle.

tube is oriented at right angles to the airflow. Air that flows into the pitot tube is brought to rest. The pressure that builds up in the tube is referred to as the total, or *stagnation,* pressure since the air in the tube is no longer moving. From Bernoulli's equation (2.8) and the fact that the air velocity in the tube is zero, the pressure in the tube P_2 is

$$P_2 = P_1 + \rho \frac{V_1^2}{2} \tag{2.10}$$

where P_1 and V_1 refer to the upstream conditions. The tube at right angles to the flow measures the static pressure, P_1. Thus the differential pressure transmitter actually senses the difference between P_2 and P_1. From (2.10) we can therefore calculate the velocity to be

$$V_1 = \sqrt{\frac{2(P_2 - P_1)}{\rho}} \tag{2.11}$$

The volumetric flow rate Q can be determined from (2.11) by multiplying the velocity by the cross-sectional area of the duct:

$$Q = A_1 V_1 = A_1 \sqrt{\frac{2(P_2 - P_1)}{\rho}} \tag{2.12}$$

The constants in (2.11) or (2.12) can be gathered up so that, in the computer, the process of finding the air velocity or the flow rate reduces to simply calculating the square root of the differential pressure input and multiplying it by the constant.

In practice, the pitot tube principle is usually implemented using perforated tubing, arrayed in the form of a circle or cross that spans the ductwork. The effect of such arrangements is to provide a stagnation pressure value that more nearly corresponds to the average reading that would be obtained with an array of individual pitot tubes, but at far less cost since only a single differential pressure sensor is required.

2.1.3.2 Orifice Plates Orifice plates also depend on a differential pressure measurement for their operation. An orifice plate is nothing more than a disk with a hole in the middle that is bolted between two pipe flanges. See Fig. 2.12. The plate introduces a pressure drop whose magnitude depends on the geometry of the plate, i.e., the ratio of the diameter of the orifice to the diameter of the pipe, the characteristics of the fluid as represented by its Reynolds number, and, of course, its velocity. (The Reynolds number of a flowing liquid or gas depends on the flow geometry, the flow velocity, and the fluid viscosity. High Reynolds numbers indicate turbulent flows where the fluid viscosity plays a minor role relative to the effect of velocity and the assumption of frictionless flow is a good one.)

As is so often the case in practice, the theory reduces to a correction factor which takes into account the deviation of the flow from that of an "ideal" fluid. These factors are derived from empirical measurements and can be found in vendors' specification sheets. The flow rate can then be expressed:

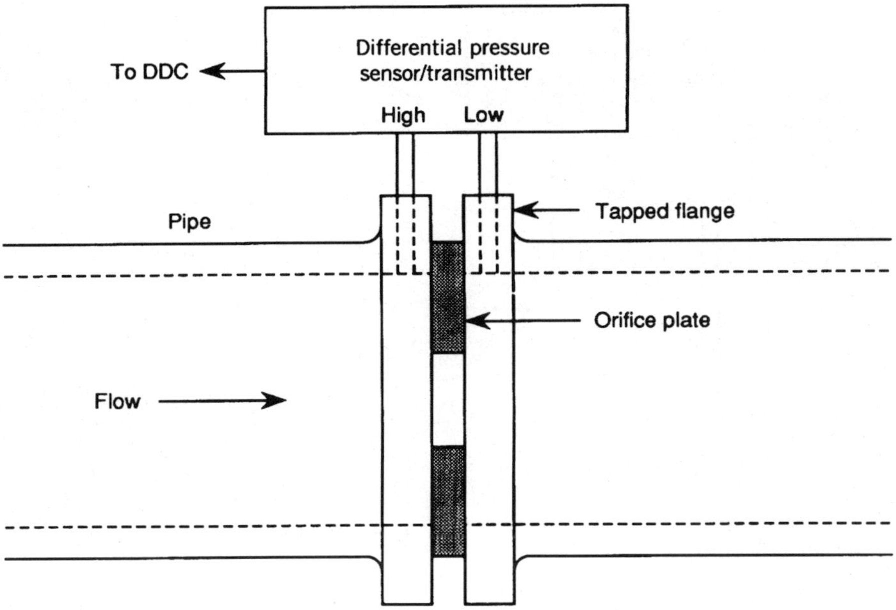

FIGURE 2.12. An orifice plate is fitted between two tapped pipes.

$$Q = CA_1V_1 = CA_1\sqrt{P_2 - P_1} = K\sqrt{\Delta P} \tag{2.13}$$

where C and K are just different forms of the correction factors.

While (2.13) can be used as is for liquids such as chilled water, gas measurements, e.g., steam flow, almost alway require taking into account the variation in the fluid's density due to temperature. Fortunately, this is where computers shine! The temperature can be easily measured and the result used to look up the correction factor corresponding to that temperature value. Alternatively, an analytic expression for the correction factor may be available as the basis for calculating, rather than looking up, the appropriate correction factor. Some vendors even supply a transmitter which does the square root extraction and temperature compensation for you so that all that is left is to read the analog input signal.

2.1.3.3 Venturi Tubes A Venturi tube, named for the late eighteenth-century Italian physicist G. B. Venturi, is a section of pipe whose diameter decreases from that of the preceding pipe to some minimum value and then increases back up again. Pressure taps are installed at the entrance to the tube and at the narrowest section. See Fig. 2.13. From Bernoulli we know that

$$P_2 + \rho\frac{V_2^2}{2} = P_1 + \rho\frac{V_1^2}{2} \tag{2.14}$$

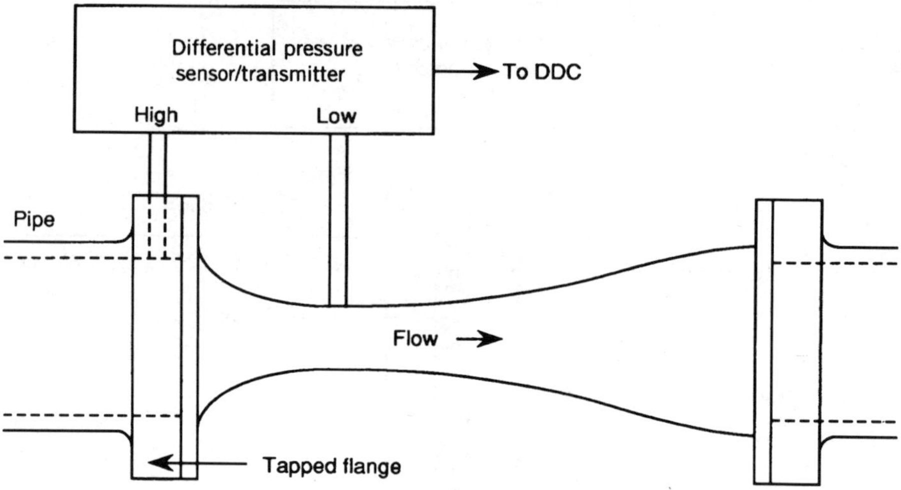

FIGURE 2.13. Because of the smooth curves of a Venturi tube, almost all of the pressure drop across the tube is recoverable.

Also, from the continuity equation, we have

$$A_2V_2 = A_1V_1 \text{ or } V_2 = V_1(A_1/A_2) \tag{2.15}$$

Substituting (2.15) into (2.14) we get

$$V_1 = \sqrt{\frac{2(P_2 - P_1)}{\rho[1 - (A_1/A_2)^2]}} \tag{2.16}$$

Again, by collecting the various constants together, the computer can be asked to process a simple expression such as (2.13).

Venturis have several advantages over orifice plates but are more expensive. The advantages are that almost all of the pressure drop across a venturi is recoverable, and venturis are not as susceptible to wear. Both of these advantages stem from the fact that an orifice plate presents a relatively brutal impediment to the fluid's flow and therefore introduces a significant disruption. The venturi, on the other hand, is far less disruptive since the pipe's diameter changes much more gradually. The bottom line is that orifice plates should be checked for wear occasionally since even a small change in the orifice diameter can lead to significant inaccuracy.

2.1.3.4 Hot-Wire Anemometers An increasingly popular alternative to pitot tube arrays for measuring airflow makes use of constant temperature thermal anemometry. (*Anemometry* is the measurement of the characteristics of the wind, from the Greek *anemos* meaning wind.) The basic idea is to insert a heated wire into the flow and measure the electrical current required to keep it at a constant temperature. See Fig. 2.14. The temperature of the heated element is chosen to be above that of the air flowing by so that a nonzero ΔT exists. To compensate for changes in the temperature of the air and to maintain a constant ΔT, most commercial anemometers have two temperature-sensing elements, one to measure the temperature of the heated element and another to measure the temperature of the airstream itself. As air passes the heated element, heat energy is transferred to the passing molecules at a rate proportional to the mass density of the air. Thus the hot-wire anemometer actually provides a measurement of mass flow which can readily be converted to standard cubic feet per minute (cfm referenced to 21°C and 1 atm). Knowing the mass density, temperature, and thermal properties of the fluid, this technique can be extended to fluids other than air with appropriately encapsulated temperature sensors and heating elements.

For airflow measurements, hot-wire anemometers have at least two advantages over pitot tube arrays. The first is that pitot tubes are susceptible to clogging with dust and dirt. The same contaminants have little effect on a hot-wire anemometer, affecting at most the device's response time to changing temperature or velocity but not the steady-state measurement itself. Second, pitot tube arrays, particularly of the perforated tube variety, are subject to errors resulting from nonuniform flow. In such cases the stagnation pressure of the air can vary from one section of the array to another with the result that there can be airflow *within* the array itself, thus violating the pitot tube's basic physical principle. Although this effect is difficult to predict or compensate for, it can be minimized by placing the array in a straight section of duct, free of obstruc-

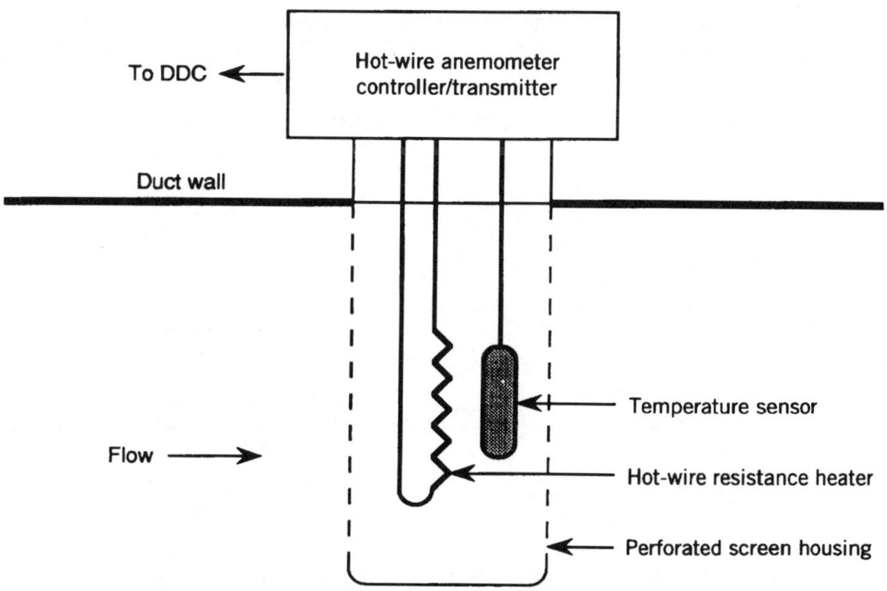

FIGURE 2.14. Hot wire anemometers provide a measurement of air mass flow.

tions, or the use of flow straighteners. A better solution is probably the use of a hot-wire anemometer *array*, a number of which are now commercially available.

2.1.3.5 Turbine Flow Meters Turbine flow meters contain a bladed rotor whose rotation speed is proportional to the volume flow rate of the fluid in which it is immersed. The rotation speed is usually sensed by a magnetic detector, which is able to detect the presence of a passing rotor blade by measuring the variations in the current required to maintain a magnetic field in the vicinity of the detector. See Fig. 2.15. The resultant signal can be processed to produce a pulse train whose frequency is proportional to the turbine's speed and, with an appropriate calibration factor, the fluid flow rate. Turbine meters are most commonly used in HVAC for measuring chilled water and steam flow, the latter requiring temperature and pressure compensation. Most turbines meters are designed for use in pipe and can either use "full bore" rotors, in which case the device occupies a pipe section mounted between flanges, or rotors which are inserted through the side of the pipe through fittings which are welded onto the pipe. These fittings can sometimes be attached to the pipe by a "hot tapping" process which allows the meter to be installed without shutting down the steam or water flow.

The biggest problem with turbine flow meters stems from the fact that they have moving parts. The rotor bearings can wear or the rotor blades can be destroyed by impurities in the flow. Think of what a rotor designed for steam service would have to endure if a slug of water, traveling at 100 ft/s or so, comes down the pipe due to an overly precipitous startup procedure!

2.1.3.6 Vortex Shedding Meters The next time there is a windy day, go find a flag and watch how it flaps! If the flag is flying off a typical sort of round flagpole, you will notice a regularity in the ripples of the cloth and the oscillation of the flag's trailing edge. It turns out that the spacing of the ripples, and thus the frequency of the flapping, is proportional to the wind speed. The ripples are actually caused by vortices

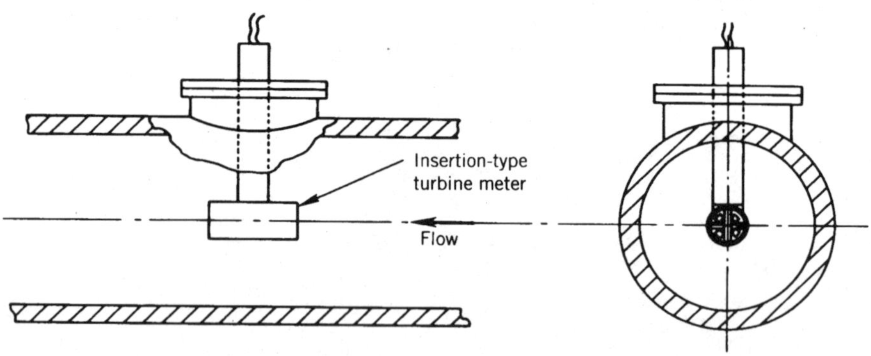

FIGURE 2.15. An insertion-type turbine flow meter. This type of device is commonly used for steam flow measurement and with the use of special tools may possibly be installed without the need for a shutdown, (NIST 1984).

which form as the air passes by the impediment to the flow represented by the flag pole. The idea of putting a "bluff body" in a fluid flow and sensing the presence of the resultant vortices is the basis of the vortex shedding flow meter.

In practice, vortex shedding meters consist of a bluff body, its squared-off side facing upstream, and a detector that can sense the pressure fluctuations caused by the vortices. See Fig. 2.16. The detector, usually consisting of strain gauges, may be physically attached to the bluff body itself or to a vane, blade, or airfoil section located a short distance downstream. In either case, the vortices produce a sinusoidal signal, proportional to flow rate, which can be manipulated to produce a pulse train or standard analog output.

2.1.3.7 Ultrasonic Flow Meters Ultrasonic flow meters use high-frequency sound waves to measure liquid flow in pipes. There are two basic types: those that measure frequency shifts due to the Doppler effect and those that measure variations in the transit time of ultrasonic sound pulses. Both types of meters have the definite advantage that they are nonintrusive; they can be bolted onto the outside of a pipe. See Fig. 2.17.

The Doppler effect, first described by Christian Doppler in the early 1800s, says that a sound wave's frequency will be increased or decreased by an amount proportional to the relative speed of the sound's source and receiver. If the source is moving toward the receiver, the perceived pitch or frequency of the sound will be higher; if the source is moving away, the frequency will be lower. The magnitude of the effect

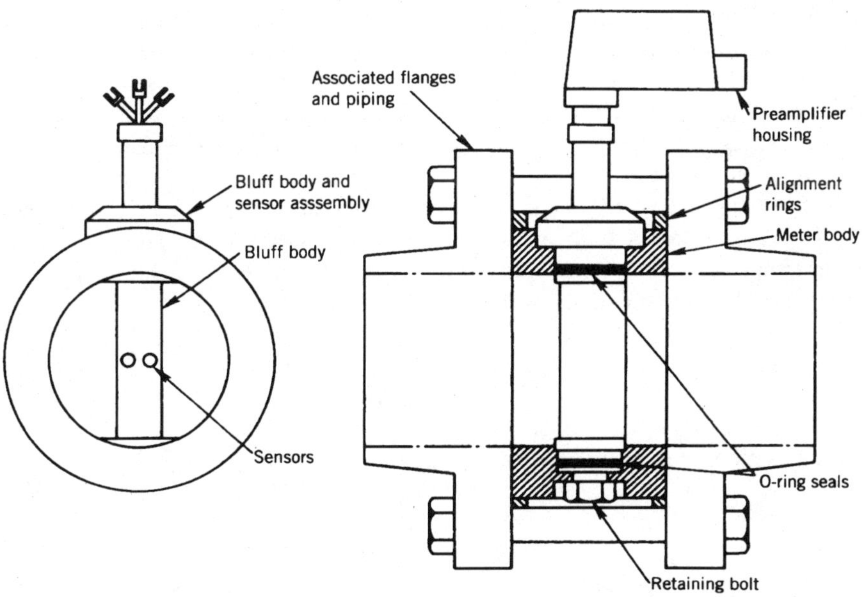

FIGURE 2.16. A vortex shedding meter. In this case, the vortex sensors are directly attached to the "bluff body" (NIST 1984).

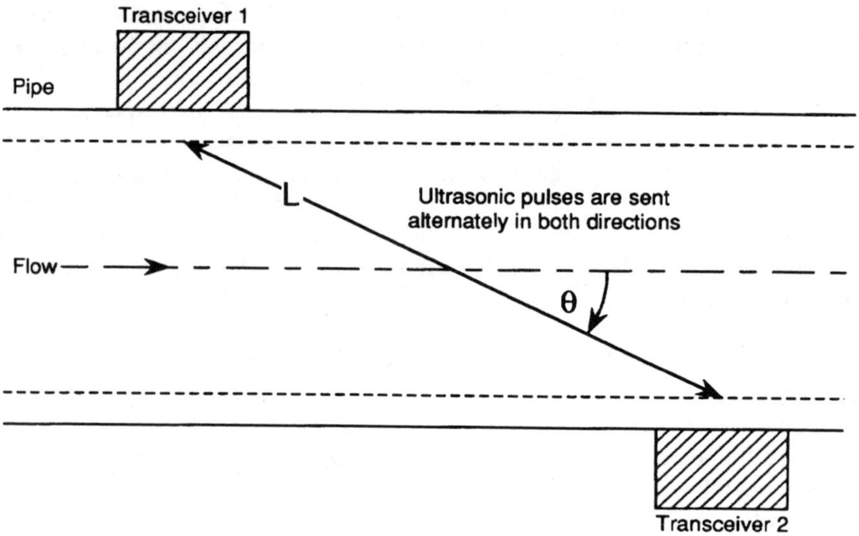

FIGURE 2.17. An ultrasonic flow meter designed to transmit pulses both up- and downstream. By calculating the difference in transit times, the velocity of the fluid may be inferred. Calibration requires knowledge of the thickness and composition of the pipe.

is related to the speed of sound in the fluid and the relative speed of the source and receiver. Practical flow meters based on the Doppler effect inject a continuous ultrasonic signal, typically of about 500 kHz, into the pipe and measure the resultant sonic reflections. The magnitude of the maximum frequency differential, $|\Delta F|$, is proportional to the maximum fluid velocity. The problem with ultrasonic flow meters based on Doppler, however, is that they require the presence of a consistent density of acoustically reflective particles, i.e., junk in the water! If the liquid is too clean, the meter may not work well.

The second type of ultrasonic flow meter relies on a measurement of the *times of flight* of sound pulses from a transmitter located upstream to a receiver located downstream, and vice versa. The time interval for a downstream pulse can be represented by the equation:

$$t_d = \frac{L}{c + k\bar{v}\cos\theta} \tag{2.17}$$

while the time interval for an upstream pulse is

$$t_u = \frac{L}{c - k\bar{v}\cos\theta} \tag{2.18}$$

In these equations, L is the distance between the transmitter and receiver; c is the speed of sound in the liquid; k is a factor that takes into account the fact that the average velocity of the flow varies along the transit path in a way that depends on fluid viscosity, pipe diameter, and maximum velocity ($k = 1$ for a perfectly uniform flow); $\bar{v}$ is the average velocity over the pipe's cross section; and θ is the angle between the transit path and the axis of the pipe. Subtracting (2.17) from (2.18) we get

$$\Delta t = t_u - t_d \cong \frac{2Lk\bar{v} \cos \theta}{c^2} \tag{2.19}$$

Equation (2.19) holds for the case where $c >> \bar{v}$. In practice, the transit time ultrasonic flow meter has its own microprocessor which is set up to solve Eq. (2.19) for $\bar{v}$. The values for L, k, θ, c, and an additional constant term used to compensate for the effect of the pipe walls on the sound propagation are derived from the manufacturer's tables for specific pipe geometries and liquids.

The main disadvantage of this type of meter is that it must be precisely matched to the specific type of pipe with which it will be used. This is only a problem if you want to move the meter around, a circumstance that might arise during commissioning or troubleshooting.

2.1.4 Humidity Sensors

Controlling humidity is another common DDC function. Although, potentially, anything whose electrical properties change in response to the amount of water vapor in the air is a candidate for use as a sensor, the most prevalent humidity sensors today are based on the use of thin-film polymers. See Table 2.5. Depending on the exact composition of the polymer, either its resistance or dielectric constant is used to provide the humidity indication. The polymer is deposited on top of a conducting grid pattern which has, in turn, been placed on an insulating substrate. The grid is composed of two halves which are electrically united by the polymer film, forming a moisture-sensitive resistor or capacitor. The entire assembly is then usually coated with a protective, but moisture-permeable, sheath. See Fig. 2.18. In laboratory tests thin-film polymer sensors have demonstrated accuracies to within 3% across the entire range of 0 to 100% relative humidity (RH).

There are two problems with thin-film polymer sensors: contamination and cali-

TABLE 2.5 Humidity Sensors

Sensor Type	Primary Use	Advantages	Disadvantages
Thin-film polymer	Relative humidity	Inexpensive	Can become contaminated with pollutants
Chilled mirror	Dewpoint temperature	Precise measurement	Mirror requires periodic cleaning, expensive

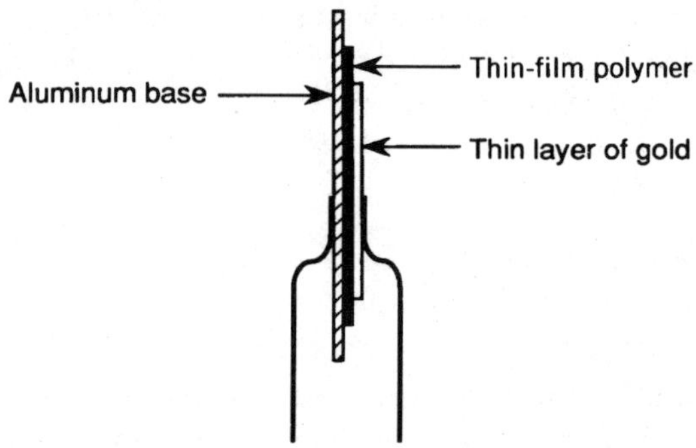

FIGURE 2.18. Schematic cross section of a typical thin-film polymer humidity sensor. As the polymer absorbs and desorbs water vapor, its dielectric constant changes resulting in a change in capacitance of the capacitor formed by the aluminum and gold plates.

bration. The first can be addressed by periodic inspection and cleaning. The second is more difficult. Probably the best approach is to obtain a high-quality portable RH meter, calibrate it using standard salt solutions available for this purpose, and then take it into the field and use its reading as a basis for adjusting the RH sensor's calibration.

Another, albeit far more costly, device for measuring humidity is the chilled mirror sensor. Developed for highly demanding industrial applications, the chilled mirror sensor actually provides a direct measurement of the dew or frost point temperature. Combined with a knowledge of the dry bulb temperature, the dewpoint temperature can be used in conjunction with the psychrometric tables to find the RH. Because of its inherent accuracy, the chilled mirror's higher cost may well be justified in certain HVAC applications such as maintaining precise humidity conditions in clean rooms, hospitals, or document repositories.

The basic operating principle is that a sample of air is passed over a mirror, the surface of which is cooled below the ambient air temperature by a small thermoelectric chiller. See Fig. 2.19. Light from a source such as a light emitting diode is reflected off the mirror to a photodetector which measures the received light intensity. When the dewpoint temperature is reached and dew appears on the mirror's surface, the amount of light reaching the detector changes abruptly. The temperature of the mirror's surface at that instant is, by definition, the dewpoint. The control circuit in the hygrometer cycles the temperature above and below the dewpoint periodically to alternately evaporate and condense the water vapor in the airstream. The obvious problem is that the mirror can become dirty over time and must be cleaned occasionally.

2.1.5 Indoor Air Quality (IAQ) Sensors

Ever since it was recognized that people can get ill breathing stale, polluted air and the term *sick building syndrome* was created to draw attention to the problem, interest in IAQ has been running high. For anyone who happens to work in a building, this is good news.

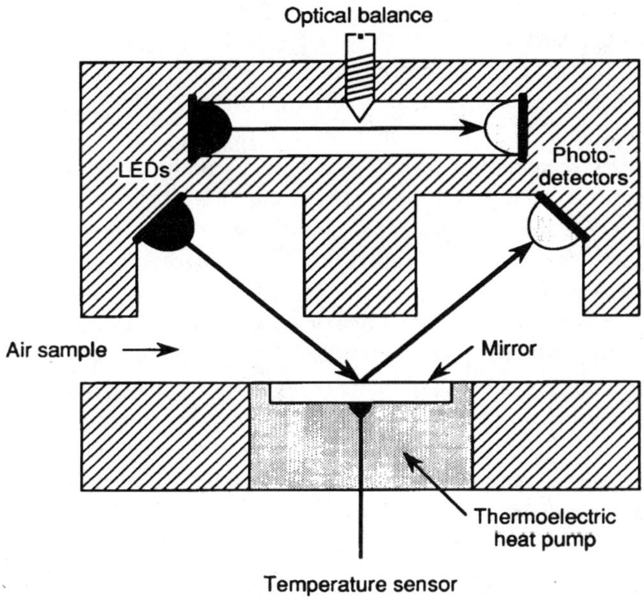

FIGURE 2.19. Schematic of a chilled mirror dewpoint sensor. Control circuitry (not shown) cycles the mirror temperature above and below the dewpoint temperature.

While there has yet to be devised a general-purpose IAQ sensor, methods exist for detecting virtually every possible contaminant. Unfortunately, most require taking a sample and performing subsequent laboratory analysis using extremely sensitive and expensive equipment such as gas or liquid chromatographs, infrared spectroscopes, colorimeters, and so on. In the absence of exotic chemicals, the sources of which must simply be isolated and mitigated, the most common contaminants are carbon dioxide and particulates such as cigarette smoke. Happily, these contaminants can be measured using continuous measurement techniques. Not only could these measured values be used as a basis for verifying that adequate ventilation exists, but they might also be able to save energy by allowing ventilation to be controlled by actual, rather than assumed, occupancy, a technique known as *demand-controlled ventilation*. This is, in fact, one of the procedures envisioned in ASHRAE Standard 62-1989, *Ventilation for Acceptable Indoor Air Quality*. In the references I also cite a 1992 paper by Drees, Wenger, and Janu which discusses airflow measuring techniques that can be used in conjunction with CO_2 sensors to comply with the standard.

2.2 SENSOR SIGNAL CONDITIONING

Since the properties of sensors that change in response to temperature, pressure, flow, or whatever else we are measuring are never standard voltages or currents, signal processing is required to convert the sensor's output into something suitable for computer input. The conditioning process generally involves amplification, linearization, and conversion of the sensor's output into a standard voltage range such as 0 to 5 VDC or the standard current range of 4 to 20 mA adopted by the ISA in their standard S50.1. (ISA 1982). The processing may be carried out at the sensor or within a field panel where several sensors may share common electronics. The latter possibility is normally only available for the most common types of sensors such as RTDs.

What are the pros and cons of doing the signal conditioning at the sensor? The main advantage is accuracy. Since the sensor output is often a very small change in voltage, current, resistance, capacitance, or inductance, the less distance it has to travel before it is measured, the less opportunity there is for the signal to be negatively affected by the environment. The disadvantages are cost, since each signal conditioner must operate as a stand-alone unit without the possibility of sharing any circuitry with other conditioners, and the need to maintain calibration at many locations rather than at just one. For many types of sensor the signal conditioning and transmitting circuitry is an integral part of what is commercially available.

If the decision is made to use remote signal conditioner/transmitters, the question arises as to whether the output of the transmitter should be a voltage or a current. This debate is usually resolved in favor of current. The reason is that what goes in one end of the wire comes out the other. Provided that the transmitter has enough power to overcome the series resistance in the loop, the current at the DDC panel will be identical to the current at the transmitter. This is not the case for a voltage signal which will inevitably suffer attenuation due to the same resistance and the fact that some small amount of current has to be drawn to permit measurement of the voltage. A disadvantage of the current loop approach is that you may have to provide a resistor across the analog input terminals at the DDC to convert the current signal back into a voltage if the DDC input is looking for a 0- to 5- or 1- to 5-V signal.

If you are wondering why the standard current range is 4 to 20 mA instead of 0 to 20 mA, consider what would happen if the transmitter should fail. If the current range began at zero, there would be no way to differentiate between the measured variable being at its low end and the absence of current because the transmitter had gone belly up. In the case of the 4- to 20-mA standard, there should always be at least 4 mA at the analog input point to the DDC. Another reason 4 to 20 was chosen is that a 250-Ω resistor across the analog input produces a signal in the 1- to 5-V range, a range compatible with the transistor-transistor-logic signal levels found in most of today's digital electronics including the analog-to-digital converters we will discuss next.

2.2.1 Multiplexing and Analog/Digital Conversion

The next link in the chain between a sensor and the use of the measured value in a calculation in the DDC computer is often a multiplexer. A *multiplexer* can be thought of as a high-speed rotary switch (without moving parts!) whose poles are connected to the incoming sensor circuits and whose output is connected to the next circuit element in the chain. In the case of analog inputs, the next link is circuitry that performs analog-to-digital (A/D, pronounced "A to D") conversion.

A/D conversion is the processing whereby an analog electrical signal is converted to a binary number. The binary number is represented by the voltage on each of several conductors called the data bus. This is represented schematically in Fig. 2.20 where the converter is taking a 0- to 5-V signal and producing a 10-bit binary number on the output lines B0 to B9. Although A/D converters were once built up from discrete components (like everything in the electronic world) and were fairly expensive, today they are available in conventional integrated circuit packages at relatively low cost. The three main characteristics of A/D converters are: (1) their input range (0 to 5 VDC is common); (2) the speed at which a signal can be converted (of great importance in certain scientific and military applications but of little concern in HVAC DDC); and (3) their output resolution.

The output resolution of an A/D converter is expressed in terms of the number of bits in the resulting binary number. In Fig. 2.20, for example, the digitized output of the 0- to 5-V input signal occupies 10 bits; hence, this figure depicts a "10-bit A/D." Since the maximum value that an N-bit binary number can assume is 2^N-1, a 10-bit A/D has a resolution of 1 part in $2^{10}-1$ or 1023; 11 bits, 1 part in 2047; 12 bits, 1 part in 4095, etc.

A practical example of the operation of a 10-bit A/D is shown in Figs. 2.21 and 2.22. Let's assume a temperature input signal in the range of 20 to 120 (in whatever units the signal conditioner has been calibrated to). At 20° the input to the A/D is exactly 0 VDC, at 120° the input is precisely 5 VDC. The A/D divides the 100°, 5 VDC span into 1023 intervals. Within each tiny interval the temperature, and hence voltage, varies but the output of the A/D *remains constant*. Each value of the binary number represents a temperature change of $100/1023 \approx 0.1°$ or a voltage change of $5/1023 \approx 0.0049$ VDC from the preceding or following binary value. Since the temperature and voltage can be anywhere in their respective intervals, there is an uncertainty in the value equal to the width of the interval.

While most general-purpose DDC panels now use at least 10 bits of A/D conversion, some smaller unitary controllers may still use only 8-bit A/Ds. This means that the span of the measured variable is broken up into just 255 parts. If this is a problem in a particular application, such as measuring a chilled water ΔT, it may be possible to reduce the span of the input signal. If the span is 100°, as in the previous example, an 8-bit A/D has a resolution of $\approx 0.4°$ but if the span is reduced to 10°, the resolution improves to 0.04°. In the worst case, it may be necessary to use a more powerful, and expensive, device.

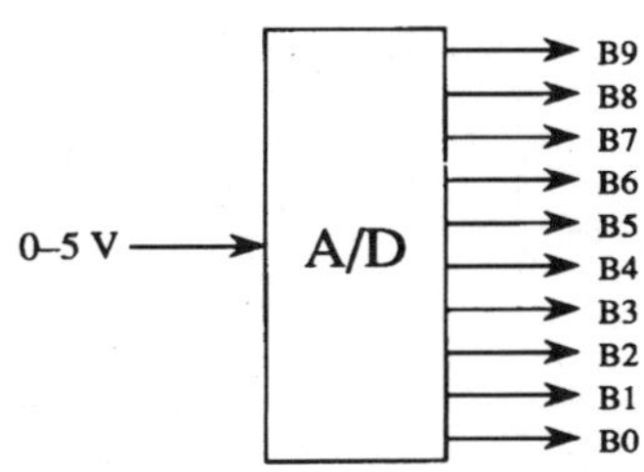

FIGURE 2.20. An A/D converter takes a standard input signal, such as 0 to 5 VDC, and converts it to a binary number. This figure shows a 10-bit A/D. The resolution of the converter depends directly on the number of bits in the resulting number.

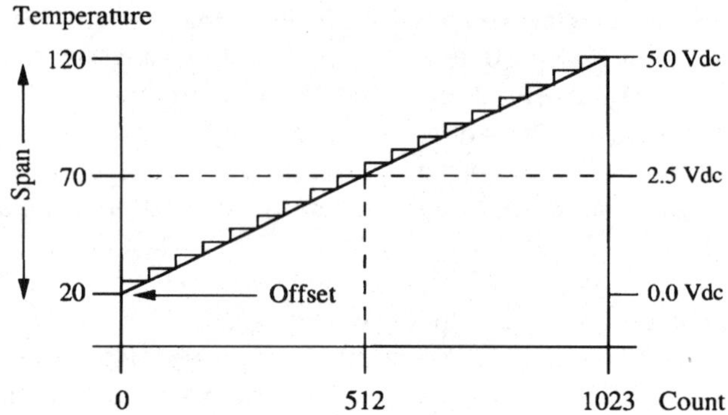

FIGURE 2.21. A 10-bit A/D converter divides the span of the measured variable into 1023 intervals. A temperature of 70° produces a voltage signal of 2.5 V which, in turn, results in a count of 512 from the A/D. Conversion constants in the computer are used to regain the sensed value, ± the resolution.

Once the digitized value of the sensor input is stored in the DDC, it may be converted into engineering units using the formula:

$$\text{VALUE} = \frac{\text{SPAN}}{\text{MAXCOUNT}} * \text{COUNT} + \text{OFFSET} \tag{2.20}$$

In the case of the example in Fig. 2.21, we have SPAN = 100, MAXCOUNT = 1023, and OFFSET = 20. A count of 512 thus corresponds to a temperature somewhere between 70.05° and 70.15°.

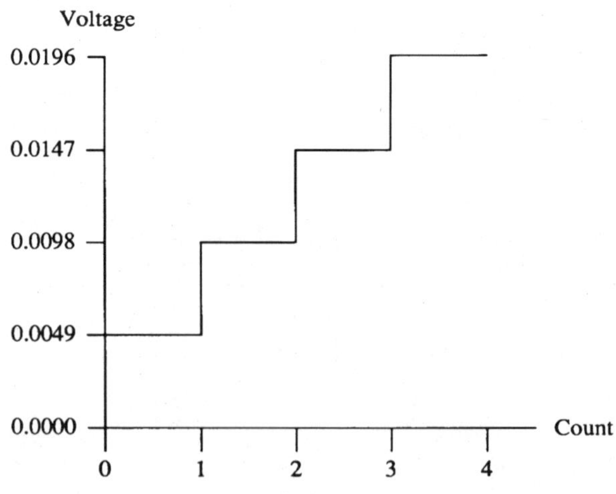

FIGURE 2.22. A close-up of the low end of the A/D curve.

2.3 SENSORS OF THE FUTURE

In looking to the future of sensors, three trends are worth noting. The first is that sensors are becoming smaller. This may ultimately translate into reduced cost since less raw material is needed to fashion these smaller devices. The second trend is the increasing development of sensors based on fiber optic technology. The third trend is toward "smart" sensors.

The trend toward reduced sensor size is a manifestation of the continuous improvements being made in microfabrication. The same techniques that led to transistors, integrated circuits, microprocessors, and now very large scale integration (VLSI) devices, are now being applied to the manufacture of sensors. Thin-film RTDs and humidity sensors that have dimensions on the order of millimeters are good examples.

Fiber optic sensors will almost certainly see much greater use in the future. Originally developed for military and aerospace applications, they are particularly well suited to applications in electrically noisy environments, such as often encountered in mechanical equipment rooms. Because they are made from nonconductive materials and transmit their signals through optical fibers, they can be placed at considerable distances from the DDC equipment to which they are attached. Since the fiber optic cable itself is made of nonconductive glass or plastic fibers, it may be possible to run it along with cables carrying electrical power, thus conserving space in ducts that carry low-voltage signal cables. See Krohn (1988) for a good overview of fiber optic sensors.

Just as information may be carried by radio waves by varying the intensity, frequency, and phase of the radio signal, so fiber optic sensors modulate the intensity, frequency, and phase of light in response to a process variable. In addition, some fiber sensors make use of the polarization of light or the travel time of light pulses. Sensors that rely on varying intensity are called *incoherent sensors* because the frequency of the light is not involved. Conversely, sensors that depend on specific wavelength- or frequency-dependent effects are called *coherent sensors.*

Many optical effects have been shown to be candidates for use in fiber sensors. It is known, for example, that glass doped with traces of rare earth oxides like neodymium absorb light in the visible and near infrared in a highly temperature-dependent way. A measurement of temperature can be derived by sending light through a sensor made of such material and comparing the intensity of the emerging light with a sample of the light that did not pass through the sensor. The ratio of intensities at particular wavelengths depends on the absorption spectrum of the sensor and can be directly related to the temperature of the sensor.

Fluorescence can also be used for temperature measurement. The rate of radiative decay of phosphors that have been bombarded by photons has been shown to depend on temperature. Temperature measurement involves sending a pulse of light from a light-emitting diode at a suitable wavelength through a fiber to a tiny fluorescent bead and then measuring the rate of fall off of the returned light.

Interferometry of coherent light sources can be used for both temperature and pressure measurements. In the former case, monochromatic light from a laser source may be sent through a fiber on the end of which is a tiny chamber filled with material whose refractive index is temperature-dependent. The end of the chamber is silvered to relect the incoming light back through the chamber and down the fiber to the source. The interference of the incident and relected beams produces interference patterns of light and dark rings, the spacing of which varys with the temperature-dependent refractive index of the material in the chamber. If the chamber is filled with nonrefractive material and the mirror on the end is allowed to flex in response to pressure changes, the same type of arrangement can be used to measure the pressure-dependent deflection of the mirror.

While the technology of fiber sensors will undoubtedly develop in parallel with advances in fiber optic communication, there are still some notable problems. One is the difficulty in multiplexing the optical signals so as to be able to use the same electronics for multiple sensors. A second is the lack of any sensor standardization with the usual result that interchangeability is difficult, if not impossible. A third problem is the lack of electrical power at the sensor, a circumstance that otherwise is viewed as a distinct advantage. The lack of power means that it is not possible to locate any computer intelligence at the sensor itself for the purpose of local signal conditioning or communication.

It is these latter capabilities that characterize the third trend I mentioned earlier: the increasing acceptance of the concept of smart sensors. With signal conditioning and communication capabilities at the sensor, the days of the 4- to 20-mA transmitter may be numbered. Instead of a current loop signal running back to centralized electronics, the digitized value of the sensor input could be transmitted on a local network, perhaps consisting of no more than a pair of wires. This, indeed, in the concept behind ISA's Fieldbus standard (SP 50) or the European Process Fieldbus (Profibus) work that we will discuss in Chapter 6. The end result would be reduced wiring costs and an increased ability to coordinate information from geographically separate locations.

2.4 ACTUATORS AND OTHER OUTPUT DEVICES

Since the purpose of DDC is control, we have to be able to interface the computer to output devices. As with sensors, we can divide the problem into binary and analog. While most binary output devices are actuated electrically, most analog output devices are still powered using pneumatic air pressure. You might wonder if this nonelectric kind of actuation is really consistent with DDC. In fact it is. Remember that the central precept of DDC is that a *computer* is determining the operation of the output device, regardless of the origin of the power producing the motion of the device. Because of the variety, widespread availability, and low cost of pneumatic actuators and the fact that most existing pre-DDC buildings already have pneumatic air compressors, pneumatics will undoubtedly be around for many years to come. Pneumatic controls, on the other hand, are likely to significantly diminish in use, if not vanish entirely!

2.4.1 Binary Output Devices

In HVAC applications, binary output signals are most commonly used to control the motors used to drive fans and pumps. Since the binary signal from the computer is likely to be a computer logic level, e.g., 0 or 5 V, mechanical relays are typically used to convert the voltage signal to a contact closure. Because DDC relays themselves are normally only rated for low-voltage applications, perhaps up to 110 V at 15 A, heavy-duty contactors are needed between the DDC relay and the actual motor, assuming it is larger than a fractional horsepower device. Contactors can be used that have extra contacts which can be wired back to a binary input on the DDC, thus providing a degree of feedback. Bear in mind that such feedback only indicates that the contactor has switched, not that the motor itself, or indeed the fan or pump, is actually operating correctly.

As for the relays themselves, there are two basic types: those that need continuous voltage to remain energized and those that toggle based on a momentary voltage pulse, thereafter remaining in whatever state they are in. Momentary pulse, or latching, relays are inherently bipolar: either they have two solenoid coils, one for each direction of travel, or they require a positive voltage pulse for travel in one direction and a negative pulse for travel in the other. The latter type have a permanent magnet attached to the switching part of the relay whose field assists or opposes that of the actuating solenoid.

The selection of a particular relay is driven by an important question: what will be the state of the relay, and hence the device to which it is connected, in the event of (1) a power failure of the DDC or (2) a software download or processor restart? Different DDC equipment behaves differently so it is important to check with the vendor or, better yet, specify the required action ahead of time.

Implementing binary control logic is one of the true pleasures of DDC. Once the binary inputs and outputs are brought back to the DDC panel, the ladder logic diagrams traditionally used to represent electrical control systems, and inflexibly hard-wired, can be replaced by software functions using Boolean ANDs and ORs that perform exactly the same function. Contact closures in series can be represented by AND: switch 1 AND switch 2 must be closed in order for the output to be energized. Similarly, parallel contacts are implemented by OR. If logic changes are needed later on, they can be made from a workstation, rather than in the field!

The mention of ladder logic immediately brings to mind the subject of programmable logic controllers (PLCs). People sometimes ask whether PLCs can be used to implement HVAC DDC strategies. (The people doing the asking are usually industrial process control engineers who are getting involved with HVAC control for the first time and have lots of experience with PLCs). The answer is a definite "yes" although some applications may be difficult. Historically, PLCs have only had limited analog capabilities so it would be important to find a PLC with enough of the right kind of analog input/output (I/O). Another, and probably more significant, problem is that PLCs are not supplied with the types of built-in library functions that typify commercial DDC gear. For example, if you needed to do humidity control you would probably have to load in the psychrometric table data or humidity computation algorithms yourself. Similarly, algorithms like optimal start/stop, which are "standard" with many vendors' DDC devices, would have to be programmed from first principles.

2.4.2 Analog Output Devices

The primary analog output devices in HVAC are valves, dampers, and variable-speed motor drives. The valves and dampers are usually powered by pneumatic actuators.

Two forms of interfacing equipment are commonplace. The first takes the numerical computer output and converts it directly into an analog voltage, current, or even pneumatic pressure. The second relies on the duration of binary pulses to generate an analog signal. This type of signal conversion is called *pulse width modulation.*

2.4.2.1 Digital/Analog Conversion Digital-to-analog (D/A) conversion is the exact counterpart of the A/D conversion discussed earlier. The idea is to take a numerical value stored in the computer and convert it to an equivalent analog signal which can then be used, ultimately, to control the movement, or in the case of a variable-speed drive, the speed of an output device. The same "stair step" behavior we discussed for A/D conversion is also present for D/A conversion. There is a definite limit to the resolution of the output signal that can be achieved for a given number of bits in the input number. Interestingly, resolution with respect to output devices is not as big an issue, at least for HVAC applications, as it is with sensors. In part this is due to the inherent "sloppiness" of the output devices themselves and the variability of the process being controlled. For a reasonably sized 3-in. water valve, for example, the difference in flow between the valve being open 50 and 55% would be difficult to pick out of the fluctuations due to turbulence, changes in supply pressure, and so on.

2.4.2.2 Analog Output Transducers: V/P, I/P Since most valves and dampers are still actuated pneumatically, it is necessary to convert the DDC's electrical signal to the pneumatic pressure signal needed to drive the valve actuator or damper motor. If the DDC's output has already gone through a D/A process and is in the form of a voltage or current, the output can be directly connected to a V/P or I/P transducer as illustrated in Figs. 2.23 and 2.24. These figures show schematically two of the most common types of electric to pneumatic transducers. In both cases the electrical signal is used as a setpoint to a comparator whose second input is a feedback signal proportional to the output pressure. In Fig. 2.23 the comparator controls a pair of two-position V/P solenoid valves. If the pressure in the central plenum chamber is below setpoint, the comparator opens the valve to the main supply air. If the pressure in the plenum is too high, the comparator opens the solenoid vented to the exhaust side. Note that no air is used when the transducer is at setpoint. Frequently the pressure signal developed in the plenum is used to drive a second transducer stage which acts as a one-to-one pneumatic relay but is capable of moving more air volume than can pass through the (usually small) solenoid valves.

In Fig. 2.24 the comparator drives a small DC motor which regulates the output pressure. The motor may actually rotate a cam which adjusts a flapper/nozzle arrangement similiar to that found in pneumatic thermostats. The pressure at the bleed nozzle is used to "pilot" a higher volume relay which delivers air to the final output device. As the cam rotates under the direction of the comparator, the flapper acts to either close off or move away from the nozzle orifice, thus modulating the back pressure to the relay.

Yet another type of I/P passes the control current directly through a wire coil suspended in the field of a permanent magnet. The resulting magnetic force balance

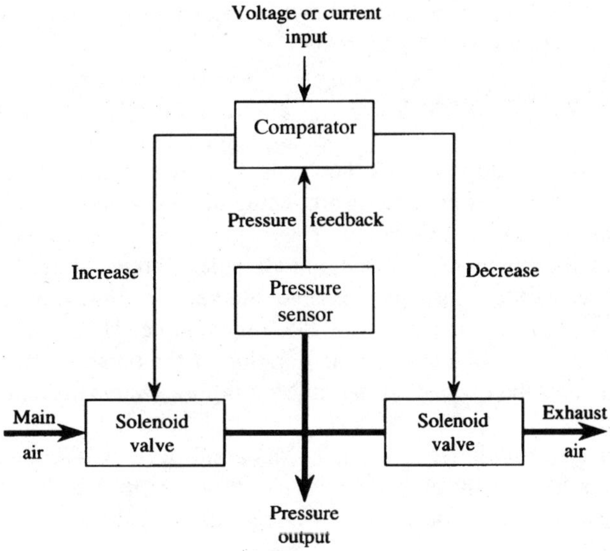

FIGURE 2.23. An V/P or I/P transducer using a pair of solenoid valves. The pressure output often pilots a 1:1 pneumatic relay for applications requiring greater air volume.

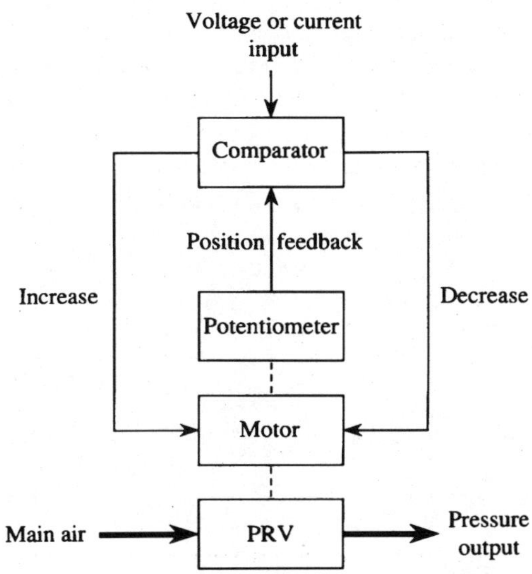

FIGURE 2.24. Another type of V/P or I/P transducer. The PRV assembly may actually be a flapper/nozzle arrangement that, again, is used to pilot a 1:1 pneumatic relay. In this device, the feedback is actually a position, rather than pressure, signal.

determines the position of a needle valve which, again, develops a pilot pressure used to drive a larger volume pneumatic relay.

2.4.2.3 Pulse Width Modulation Primarily for reasons of economics, many of the earliest DDCs did not provide true analog output signals. These devices were still capable of driving analog devices, however, by pulsing their binary outputs in a precisely timed way. This method of producing an analog signal is generally called pulse width modulation, or PWM.

Several interface methods are used, but all depend on the ability of the computer to turn on a binary output for a precise time interval. The time interval is referenced to an integer count, in the range of 1 to 255, for example. The count is multiplied by a fixed time increment to determine the duration of the binary pulse. Depending on the application and the controller, the time multiplier increment may only be a few milliseconds.

Consider a time multiplier of 20 ms. The output pulse would then vary between 0.02 and 5.10 s for a count of 1 to 255. This pulse can then be fed to a circuit that detects the rising edge of the pulse, measures the time duration, and develops an analog output signal proportional to the length of the pulse. In the example given, a 0.02-s pulse might result in a 4-mA current output, a 5.1-s pulse in a 20-mA output.

Another common arrangement is to use a *pair* of binary outputs to bypass the comparator stage shown in Figs. 2.23 and 2.24, thus driving the solenoids or pressure-regulating motor directly. The feedback signal can then be taken directly back to the DDC which performs the function of the comparator in software.

In some DDCs, binary outputs are grouped in pairs such that one output is either at ground potential or a positive voltage while the other is at ground or a negative voltage. By using the outputs together, a tri-state output is available which can correspond to a signed integer value in the computer, e.g., −255 to +255. This arrangement can also be used to generate an analog output by passing the tri-state PWM signal through a converter as shown in Fig. 2.25.

While the techniques just discussed presumed that a true analog signal would be developed corresponding to a specific position or speed of the output device, PWM

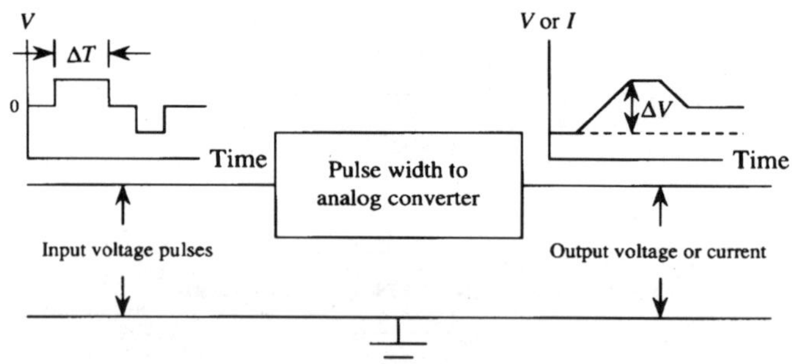

FIGURE 2.25. A pulse width to analog converter where the input is bipolar, i.e., swings positive and negative with respect to circuit ground. Converters are also available for unipolar inputs.

also lends itself very well to incremental control where *changes* in the output device's position or speed are made in lieu of trying to establish a specific position. This technique is often used where the only feedback available is from the process itself.

In the end, all of the methods described produce pretty much the same result. The differences between them are found in the cost of the intervening transducers and personal preference. I prefer the slightly more costly positional control strategies because you can actually have the computer tell you where the valve is supposed to be, rather than how long a pulse was last used to move it in a given direction!

REFERENCES

ANSI/MC 96.1-1982, *Temperature Measurement Thermocouples,* New York: ANSI, 1982.

ASHRAE Standard 62-1989, *Ventilation for Acceptable Indoor Air Quality,* Atlanta, GA: ASHRAE, 1989.

Benedict, R. P., *Fundamentals of Temperature, Pressure, and Flow Measurements*, 3rd ed., New York: Wiley, 1984.

Drees, K. H., Wenger, J. D., and Janu, G., Ventilation Air Flow Measurement for ASHRAE Standard 62-1989, *ASHRAE Journal*, October Vol. 34, No. 10, 40–45, 1992.

ISA Standard S50.1, *Compatibility of Analog Signals for Electronic Industrial Process Instruments,* Research Triangle Park, NC: Instrument Society of America, 1982.

Krohn, D. A., *Fiber Optic Sensors,* Research Triangle Park, NC: Instrument Society of America, 1988.

BIBLIOGRAPHY

Allocca, J. A., and Stuart, A., *Transducers—Theory and Applications,* Reston, VA: Reston Publishing Company, 1984.

Baker, D. W., and Hurley, C. W., *On-Site Calibration of Flow Metering Systems Installed in Buildings,* NBS Building Science Series 159, Washington, DC: National Institute of Standards and Technology (formerly the National Bureau of Standards), 1984.

Chester, A. N., et al., *Optical Fiber Sensors*, Boston, Dordrecht, Lancaster: Martinus Nijhoff Publishers, 1987. A collection of papers edited by Chester.

Hordeski, M. F., *The Design of Microprocessor, Sensor, and Control Systems*, Reston, VA: Reston Publishing Company, 1985.

Omega Engineering, *The Flow and Level Handbook*, Stamford, CT: Omega Engineering, Inc., 1993a.

Omega Engineering, *The Pressure Strain and Force Handbook*, Stamford, CT: Omega Engineering, Inc., 1993b.

Omega Engineering, *The Temperature Handbook*, Stamford, CT: Omega Engineering, Inc., 1993c.

Smith, E., *Principles of Industrial Measurement for Control Applications*, Research Triangle Park, NC: Instrument Society of America, 1984.

Chapter 3

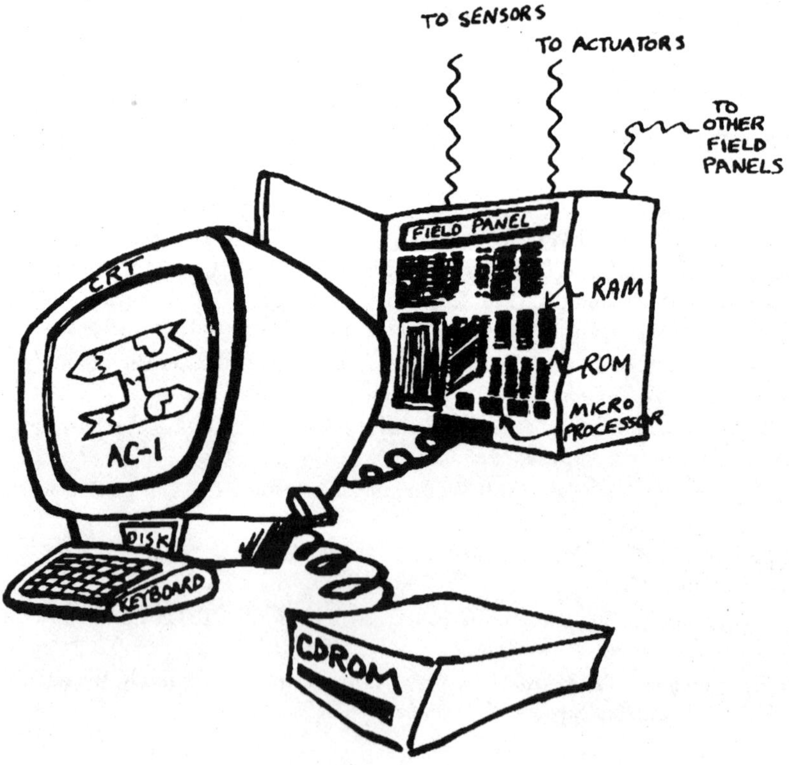
TO SENSORS
TO ACTUATORS
TO OTHER FIELD PANELS
FIELD PANEL
RAM
ROM
MICRO PROCESSOR
CRT
AC-1
DISK
KEYBOARD
CDROM

DDC Hardware

SUMMARY

It is always astonishing to me to reflect that all the electronic, digital computing that has been done by the human race has been performed since 1943! Spurred on by World War II, mathematicians, scientists, and engineers worked feverishly to develop machines capable of decrypting the enemies' codes. The result were machines like the British Colussus I and the Harvard Mark I. The Colossus, in particular, played a pivotal role in the defeat of the Nazis. Three years later, after hostilities had ceased, the Electronic Numerical Integrator and Calculator, the ENIAC, ushered in what most now refer to as the "computer age."

It is also amazing to consider that the now ubiquitous personal desktop computers have only been with us in force since the mid-eighties, only a few years after the first real DDC systems were installed in any great numbers. Yet in spite of the massive proliferation of computers, the great improvements in performance, and the dramatic reductions in size and cost, the basic principles of operation have changed but little!

Fundamentally, all computers (with the exception of the new parallel processor machines) work the same way, sequentially executing series of instructions that they retrieve from one of several possible storage areas: RAM, ROM, or disk. The instructions direct the processor either to act on data or components of the machine itself. The machine components are linked by common multiconductor address, data, and control buses. The function of the address and data buses depends on the signals present on the control bus, e.g., an address can refer to a memory location or a register in the controller of a peripheral device. Similarly, these signals determine where the data on the data bus ends up at the conclusion of a particular machine cycle.

Developments in miniaturization have led to the consolidation of computing functions on single integrated circuit chips. This very large scale integration (VLSI) technology has led, in turn, to the development of *application specific integrated circuits* (ASICs), or *microcontrollers,* which combine processors, memory, timers, counters, and input/output capabilities in one package.

For the HVAC industry, these advances in computer technology mean that it is now possible to provide sophisticated, cost-effective computer control for virtually any device, either stand-alone or as a part of a larger, more comprehensive network of devices.

Computer hardware, as applied to DDC, can be studied from three different views. The first view is from the outside: how do DDC computers interact with the external world? We have discussed this view at some length in the previous chapter when we discussed multiplexers, A/D conversion, and so on. In this chapter we will consider the second view: from the inside. How do DDC computers actually work at the machine instruction level. This is basic material with which everyone who works with DDC should at least be familiar. Finally we will look at packaging: what kind of DDC devices are available in the marketplace and what determines their configuration?

3.1 FUNDAMENTALS

All DDC systems consist of hardware, software, and firmware. The *hardware* is the physical machinery consisting of the microprocessor(s), memory, and various peripheral devices such as the sensor/actuator interfaces, serial and parallel communication ports, and operator interfaces. The peripherals communicate with the HVAC sensors and actuators, other computers, and humans. The *software* is the set of instructions and procedures that tells the computer what computations to perform and what to do with the results. *Firmware* refers to software instructions that are stored in silicon, on hardware memory chips. Firmware thus has some of the attributes of both hardware and software: it is a set of software instructions that can be plugged into an integrated circuit socket!

There are several levels of instructions or *programs* that we will discuss in the coming pages. In all cases, however, the information is stored and processed in *binary* format. For this reason, some understanding of how data are represented and stored is a prerequisite to understanding how computers work.

3.2 DATA REPRESENTATION

As just noted, *all* data are stored and manipulated within a DDC in the form of 0's and 1's. This two-state format is used because of the ease of fabricating two-state electronic devices, i.e., devices that are in one of two possible conditions: conducting/nonconducting, magnetized/demagnetized, positive polarity/negative polarity, etc. But the same string of 0's and 1's may represent a wide variety of different pieces of information. The string '1000001', for example, could be used to represent the decimal value 65, the ON or OFF status of seven fans or pumps, or the letter 'A'. We will return to this point in a moment.

Binary numbers consist of strings of 0's and 1's, which, when multiplied by powers of 2 corresponding to the digit's position in the string, can be used to represent equivalent decimal values. There are also various codes, such as the American National Standards Institute's *American Standard Code for Information Interchange* (ASCII), which allow strings of binary digits to represent alphanumeric characters or "control" characters, such as the familiar "carriage return" or "line feed," whose use may vary from system to system. See Table 3.1. The individual *bi*nary digi*ts* are referred to as *bits*. Groups of bits are referred to as *bytes*, the most common length of which is 8 bits. Bytes are grouped together as "words." The more storage locations that can be "addressed," or accessed, with a single operation, the faster a given computation can be performed. The categorization of computers into microcomputers, minicomputers, and mainframes can be roughly correlated to the binary word length that each processor uses: microcomputers generally employ 8-, 16-, or increasingly, 32-bit words; minicomputers use 16- or 32-bit words; and mainframes use 32-bit words or larger.

Four numbering systems are frequently encountered in computer literature: binary, octal, decimal, and hexadecimal. As "positional" numbering systems, they are conceptually identical in every regard. They differ only in the value that has been selected as their *base,* or *radix*. The binary system is said to be of base 2, the octal system of base 8, the decimal system of base 10, and the hexadecimal system of base 16. Notice that the decimal system is the peculiar one: its base is the only one that is not a power of 2! See Figs. 3.1 and 3.2.

What are the common characteristics of positional numbering systems? Numbers in all positional number systems are represented by strings of digits. Each digit can take on as many different values as the value of the base. Thus in the binary system a digit can have one of 2 values, in the octal system one of 8 values, in the decimal system one of 10 values, etc. These values start at zero and increase to a maximum of the base minus 1: binary digits can be just 0 or 1, octal digits range from 0 to 7, decimal digits range from 0 to 9, and hexadecimal digits can range between 0 and 15. Since a digit must be represented by a single symbol, the values 10 through 15 in the hexadecimal system are represented by the characters A through F. The digit's position in the string is significant and its value is determined by multiplying the value of the digit by the power of the base corresponding to the digit's position. The rightmost digit occupies position zero.

Consider the decimal number 543, for example. To find out how may "things," e.g., apples, oranges, or degrees Fahrenheit this number represents, we note that 543 decimal = $(5 \times 10^2) + (4 \times 10^1) + (3 \times 10^0) = 500 + 40 + 3$. Remember that any number raised to the power of 0 is equal to 1. In a similar fashion, 101 binary = $(1 \times 2^2) + (0 \times 2^1) + (1 \times 2^0) = (4 + 0 + 1)$. So binary 101 equals decimal 5. Moreover, the number of different values that an N-digit number in a system of base B can take on is B^N. The smallest value in such a case is 0 and the largest value is $(B^N - 1)$. For example, a 3-digit decimal number can take on $10^3 = 1000$ different values ranging from 0 to $(10^3 - 1) = 999$. Similarly, a 3-digit binary number can take on $2^3 = 8$ values

TABLE 3.1 American Standard Code for Information Interchange

B7 / B6 / B5				0 0 0	0 0 1	0 1 0	0 1 1	1 0 0	1 0 1	1 1 0	1 1 1
B4	B3	B2	B1	CONTROL		FIGURES		UPPERCASE		LOWERCASE	
0	0	0	0	NU 0	DL 16	SP 32	0 48	@ 64	P 80	` 96	p 112
0	0	0	1	SH 1	D1 17	! 33	1 49	A 65	Q 81	a 97	q 113
0	0	1	0	SX 2	D2 18	" 34	2 50	B 66	R 82	b 98	r 114
0	0	1	1	EX 3	D3 19	# 35	3 51	C 67	S 83	c 99	s 115
0	1	0	0	ET 4	D4 20	$ 36	4 52	D 68	T 84	d 100	t 116
0	1	0	1	EQ 5	NK 21	% 37	5 53	E 69	U 85	e 101	u 117
0	1	1	0	AK 6	SY 22	& 38	6 54	F 70	V 86	f 102	v 118
0	1	1	1	BL 7	EB 23	' 39	7 55	G 71	W 87	g 103	w 119
1	0	0	0	BS 8	CN 24	(40	8 56	H 72	X 88	h 104	x 120
1	0	0	1	HT 9	EM 25	) 41	9 57	I 73	Y 89	i 105	y 121
1	0	1	0	LF 10	SB 26	* 42	: 58	J 74	Z 90	j 106	z 122
1	0	1	1	VT 11	EC 27	+ 43	; 59	K 75	[91	k 107	{ 123
1	1	0	0	FF 12	FS 28	, 44	< 60	L 76	\ 92	l 108	| 124
1	1	0	1	CR 13	GS 29	- 45	= 61	M 77	] 93	m 109	} 125
1	1	1	0	SO 14	RS 30	. 46	> 62	N 78	^ 94	n 110	~ 126
1	1	1	1	SI 15	US 31	/ 47	? 63	O 79	_ 95	o 111	DT 127

Decimal number system

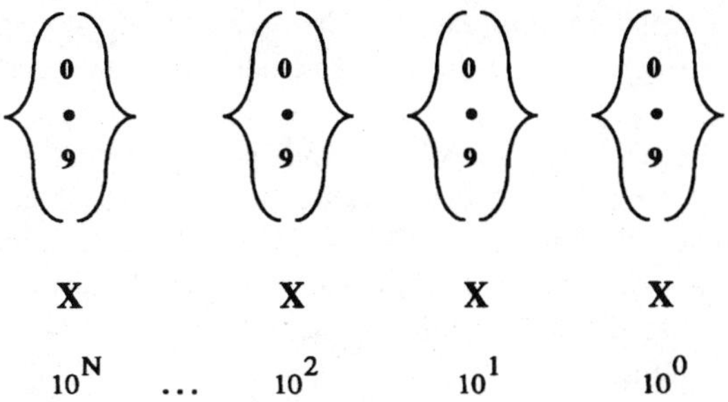

Example: $D'241' = 2 \times 10^2 + 4 \times 10^1 + 1 \times 10^0$

10^N = Number of combinations of N decimal digits

$10^N - 1$ = Largest number that can be represented by N decimal digits

FIGURE 3.1. The familiar decimal number system. Like all positional numbering systems, the value of each digit is found by multiplying its value times a power of the base corresponding to its position in the string of digits.

Binary number system

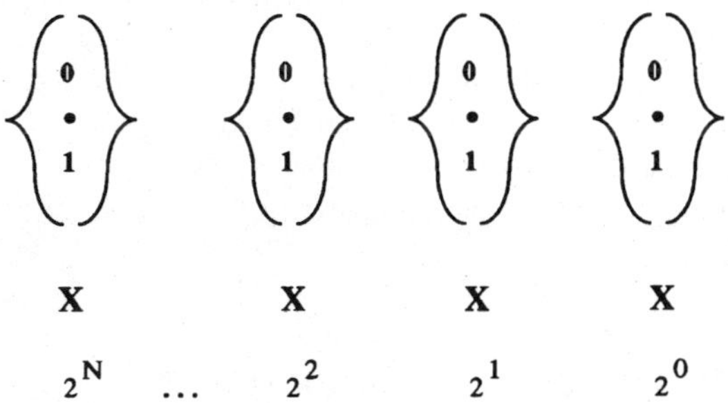

Example: $B'101' = 1 \times 2^2 + 0 \times 2^1 + 1 \times 2^0 = D'5'$

2^N = Number of combinations of N binary digits

$2^N - 1$ = Largest number that can be represented by N binary digits

FIGURE 3.2. The binary number system is identical in concept to the decimal system except that the base is 2 instead of 10.

ranging from 0 to $(2^3 - 1) = 7$. We have already made good use of these facts in our discussion of A/D and D/A conversion in the last chapter.

The binary system is important because data are actually stored and manipulated in a computer as strings of 0's and 1's. The octal and hexadecimal systems are important because they provide a shorthand way of representing groups of bits. Three bits can represent the decimal values 0 to 7 inclusively and can be represented by a single octal digit. All possible decimal values of four bits, 0 to 15, can be conveniently represented by a single hexadecimal digit, 0 to F. Since bits are usually grouped in 8-bit bytes, octal representation can be somewhat awkward. Hexadecimal is ideal, however, since all 256 possible values of an 8-bit byte can be represented exactly by two hex digits.

In addition to representing numbers, groups of bits can also represent character information such as upper and lowercase letters, punctuation marks (such as .,?,!), arithmetic symbols (such as +, −, %, *, =), and so on. This is accomplished by setting up a one-to-one correspondence between a binary number and the desired character. This correspondence is called a *code.*

If, for example, we wished to represent the 26 letters of the alphabet, we could let the number 1 represent A, 2 represent B, etc. Since a 5-bit binary number can take on 32 different values, all the letters could be represented with 6 values left over to represent other symbols of our choosing. The Baudot code of the 1960s, used by the old teletype services, was just such an arrangement. If we desired to represent a greater number of symbols, longer bit strings could be used. The ASCII code, mentioned earlier, is a 7-bit code and can therefore represent 2^7 or 128 different symbols. Not surprisingly, ASCII is by far the most common code in use today in the United States.

ASCII is, however, by its very nature severely limited. The need exists to be able to represent far more than just 128 symbols or even the 256 symbols made possible by the use of an 8-bit byte. One possibility is being developed by a consortium of U.S. companies including Aldus, Apple, IBM, Microsoft, and Xerox, among others. It is called Unicode and is a 16-bit, 65,536 character proposal. With this number of possible characters Unicode should be able to encompass all modern and obsolete languages, represent all sorts of math and technical symbols, and have room to spare for future additions.

3.3 MICROPROCESSORS/MICROCOMPUTERS/ MICROCONTROLLERS

Perhaps more than any other factor, the development of microprocessors has been responsible for the explosive growth of the computer industry. *Microprocessors* are "computers on a chip." While early microprocessors required many additional, external components in order to perform any useful work, the increasing use of large scale integration (LSI) or VLSI semiconductor fabrication techniques has led to the production of *microcomputers* where all of the required circuitry is embedded on one, or at most a small number, of integrated circuits. A further extension of this integration is the single-chip *microcontroller* which adds analog and binary I/O, timers, and counters so as to be able to carry out real-time control functions with almost no additional hardware.

At the heart of every microcomputer is circuitry to perform arithmetic and logical computations, circuitry to control the flow of data into and out of the processor, and some form of "memory" to hold the set of instructions, i.e., the program, the processor is to carry out. In the old days, when computers took up entire rooms(!), the terms *central processing unit* (CPU) and *arithmetic logic unit* (ALU) were used to describe the equipment carrying out these basic functions. These terms are still in use today, although the equipment now fits on the head of a pin.

Program instructions are normally executed sequentially. However, if certain logical conditions are met, e.g., one variable is found to be less than (or equal to or greater than) another, or the result of a computation results in zero, then the sequence of execution may be altered. This process is known as *conditional branching.* We will discuss this in more detail when we talk about instruction sets.

Special sets of instructions are given control of the machine in response to external occurrences or *interrupts* whose occurrence is not in any way synchronized with the normal operation of the computer. Interrupts may be generated by the arrival of a character from a keyboard or communication line, the closing of an external contact on some form of switching device, or other things which require the computer to take immediate notice. Data, in contrast to program instructions, which the computer uses in its tasks may also be stored in the CPU's memory, or they may be stored in some peripheral storage device outside of the CPU.

Microcomputer memory takes several forms. It may be *read only memory,* or ROM, in which case the instructions and/or data it contains are installed or "burned in" prior to installation and may not be altered under normal circumstances. This unchangeable set of program instructions is the *firmware* mentioned earlier since the *software* has been installed in a "hard" or permanent fashion.

The memory may also be *read/write memory,* often referred to as *random access memory,* or RAM. Although the memory locations in a ROM may also be accessed randomly, they cannot be changed or "written" into. An important characteristic of RAM is that it is "volatile," i.e., its contents are lost in the event of a power outage. To protect against this potential loss, most RAM-based DDC systems incorporate some form of *uninterruptible power supply* (UPS). For small systems, the UPS may be nothing more than a rechargeable battery with only enough power to sustain the RAM for some period of time, several days being typical. In large systems, the UPS may be capable of sustaining the entire system for hours, or it may be designed to provide only enough power to permit an orderly shutdown of the computer without loss of data by writing the contents of the RAM to some external mass storage device such as a disk.

Microprocessors trace their heritage back to the days just shortly after the first successful integrated circuits were produced by Fairchild and Texas Instruments in the late 1960s. The first microprocessor was the Intel 4004, a 4-bit machine produced for use in handheld calculators in 1971. The next year Intel came out with the 8008, an 8-bit machine, that required approximately 20 more ICs to form a functional computer.

In 1974 Intel introduced the 8080 whose main attractions were a larger instruction set than the 8008 and the need for only two additional chips to function. About the same time, Motorola brought out the MC6800. Its main advantage was that it only needed a single +5 VDC power supply instead of the ±5 and +12 VDC supplies needed by the 8080. Almost concurrently, MOS Technology produced a chip called the 6502 which was used in the Apple II computer, and Zilog introduced the Z80 microprocessor which found a home in the Radio Shack TRS-80. The race was on!

The "modern" era probably can be reckoned from the introduction by National Semiconductor and Texas Instruments of the 9900 16-bit microprocessor in the mid-1970s. This was quickly followed by Intel's 8086 and Motorola's MC68000, also 16-bit machines. The Intel 8088 is virtually identical to the 8086 except that it uses an 8-bit data bus. It was used in the original IBM PC and in various first generation DDCs such as the Johnson Controls DSC 8500.

Two trends in microcomputerization are relevant to DDC. The first is the ever increasing power of general-purpose microcomputers. From their humble 4-bit beginnings, microprocessors capable of processing 32-bit addresses and data are becoming increasingly commonplace. The second trend is the steady development of ever more powerful single-chip microcontrollers such as the Intel 8051 and 8096 and the Motorola MC68HC11. These chips, developed largely in response to the automotive industry's desire for computer-controlled ignition, emission control, and antiskid systems, are now widely used in unitary controllers for DDC applications such as VAV box or packaged AC unit control. Table 3.2 summarizes the evolution of the Intel 80XXX family of general-purpose microprocessors while Table 3.3 does the same for the Motorola 68XXX family. Table 3.4 compares the Intel 8051 and 8096 and Motorola 68HC11 microcontrollers.

3.4 MICROPROCESSOR PROGRAMMING

The best way to really understand the internal operation of microcomputers is to study each processor's instruction set. Although fascinating in itself, this topic is somewhat outside the bounds of the present work, so in this section we will only deal with a few of the most important aspects of it.

Most microprocessor instructions can be placed into one of five distinct categories: data transfer, arithmetic, logic, control transfer, and processor control. In addition, there may be some miscellaneous instructions for string manipulation or other specialized tasks.

Each instruction consists of an operator and zero or more operands. The operators are represented in programs as mnemonic character strings. The term *mnemonic* means the string is supposed to remind you what the function of the operator is. For example, the data transfer instruction MOV in the 8086 instruction set copies a byte or word from a specified source to a specified destination. The location specified may

TABLE 3.2 Intel 80XXX Family

8086	16-bit ALU, 16-bit data bus, 20-bit address bus. Can address 2^{20} = 1,048,576 locations.
8088	Same as 8086 except for 8-bit data bus. Used in original IBM PC and several first-generation DDCs.
80186/88	Improved versions of 8086/88. Instruction set is a superset containing 10 new instructions. Also have integrated peripherals including clock pulse generator, interrupt controller, counter/timer, and direct memory access (DMA) unit. DMA unit allows I/O <-> I/O, memory <-> memory, and I/O <-> memory transfers with minimum CPU intervention.
80286	16-bit ALU, 16-bit data bus, 24-bit address bus. Can address 16 megabytes of physical memory. Supports virtual memory operations up to 1 gigabyte and has instructions to allow multiuser, multitasking operating system. Used in the PC/AT.
80386	Full 32-bit processor that improves upon the virtual addressing mode of the 80286 by allowing smaller segments of code or data (pages) to be moved into and out of memory.
80486	168-pin software compatible but improved version of the 80386. Runs about twice as fast. Has a built-in floating point coprocessor, memory management unit, and 8 kilobytes code and data cache for prefetching of instructions and data. Some of the most commonly used instructions have been designed to execute in one clock cycle, similar to reduced instruction set computer (RISC) processors.

be a hardware register in the CPU, a register in an internal or external peripheral controller, or a location in memory.

The method by which locations are referenced, or "addressed," is the key to understanding the architecture of any particular microprocessor. The simplest method is direct addressing where the address has the same number of bits as the address bus. This is easy if the the registers and buses have the same number of bits, as is the case with the 80386, for example. In the case of the 16-bit 8086 family, the address bus is 20 bits, but all the internal registers are 16 bits. To generate the 20-bit physical address, two 16-bit registers are combined. The first contains the start of a block of memory called a *segment*. The second contains the offset of a particular byte within the segment. By shifting the segment register to the left four bits and adding the offset, a full 20-bit physical address is produced which can be output on the address bus and can thus access any of 2^{20} = 1,048,576 memory locations.

To increase the amount of memory that can be accessed, the concept of *virtual memory* has been developed. The basic idea is to shift code and data into memory from external storage when it is needed and write it back out when a task is done with it. The information in external storage is assigned a range of virtual addresses that the hardware recognizes as external. When such addresses are encountered, the hardware loads the data into physical memory for processing. Another enhancement is to assign each segment a privilege level so that only authorized tasks can gain access to it. One

TABLE 3.3 Motorola 68XXX Family

68000	16-bit ALU, 16-bit data bus, 24-bit address bus. Supports 16 megabytes of address range. Normally contained in 64-pin dual-in-line package (DIP). Versions run at 4 and 12 MHz.
68008	Reduced bus 8-bit version of the 68000 with 8-bit data bus and an effective 20-bit address bus that supports 1 megabyte of address range. Constructed in 48-pin DIP.
68010	16-bit microprocessor incorporating virtual memory. It is pin compatible with the 68000.
68012	16-bit microprocessor with enhanced virtual memory. Supports up to 2 gigabytes of memory and multiprocessing capability. Comes in an 84-pin grid-array package.
68020	32-bit cache memory processor. Cache memory holds prefetched information and thus speeds up operation. It has 32-bit data and address buses and supports 4 gigabytes of memory. It also has additional control and interface circuitry to support the 68881 floating point math coprocessor.
68030	32-bit enhanced cache memory processor. In addition to the features of the 68020, it has an internal 256-byte data cache memory and a built-in memory management unit (MMU).
68040	32-bit processor incorporating a high degree of parallel processing. Contains a built in floating point math coprocessor that can operate concurrently with integer processor. Dual MMUs allow simultaneous data and instruction access. Has two 4-kilobyte caches for data and instructions. Separate internal buses allow concurrent data movement and manipulation. At 25-MHz clock speed, can execute about 20 million instructions per second.

TABLE 3.4 Comparison of Three Microcontrollers

8051	68HC11	8096
8-bit CPU	8-bit CPU	16-bit CPU
4K ROM	8K ROM, 512-byte EEPROM	8K ROM
128-byte RAM	256-byte RAM	232-byte RAM
2 16-bit timer/counters	16-bit timer	16-bit timer
	8-bit counter	16-bit counter
Serial port	Serial port	Serial port
32 binary I/O	16 binary I/O	40 binary I/O
	8-channel, 8-bit A/D	8-channel, 10-bit A/D
	Watchdog timer	Watchdog timer
	8-bit pulse accumulator	Pulse width output

use of this is to keep undebugged user programs from inadvertently wiping out the operating system by making changes to system code or data areas. Virtual memory is the basis for most multiuser and/or multitasking operating systems in current use.

One of the most important functions of the data transfer instructions is to move data to and from peripheral devices. These devices, such as disk drives, communication ports, and the input/output boards that interface to our analog and binary sensors and actuators, are connected to the internal computer buses via I/O *controllers*. Each controller typically has registers attached to the address, control, and data buses. Each controller also has an assigned address. When the controller sees its address on the address bus, and the control bus indicates that data is to be read from a peripheral device rather than a memory location, it puts whatever data it may have in its data register onto the data lines. The process of outputting data is identical except that the data is transferred in the other direction, from the data bus to the device's data register. See Fig. 3.3. What happens next depends on the type of peripheral.

3.5 DDC PACKAGING

One of the virtues of DDC is that equipment is available with a wide range of I/O capabilities and sophistication, but all making use of the same basic principles of operation. The difference is basically one of *packaging*, how much capability is available in a specific device in terms of computing power and interfacing. The resulting variety in packaging means that it is usually possible to find a device well suited for any particular application. In this section, I would like to present an overview of the types of packaging currently available, from general-purpose DDCs to "smart" sensors and actuators.

3.5.1 General-Purpose DDCs

The main distinguishing characteristic of general-purpose DDCs is that they are field programmable. The programming techniques vary from system to system and may require knowledge of a text-based programming language or the use of some kind of computer-aided software development program. In the latter category are graphical programming tools that allow software developers to turn schematic representations of control logic directly into machine-readable instructions. Programming techniques are discussed more fully in the next chapter.

Whatever the programming technique, the general-purpose DDC can be made to carry out almost any conceivable control task. This is usually made possible by the presence of libraries of preprogrammed routines, generally installed in firmware in the form of procedures or functions that can be invoked, or "called," from user programs. Such routines are provided to facilitate performing the most common HVAC control functions. A good example would be functions to perform closed-loop, feedback control, perhaps based on a proportional-integral-derivative (PID) calculation. The programmer simply has to supply the names of the control and

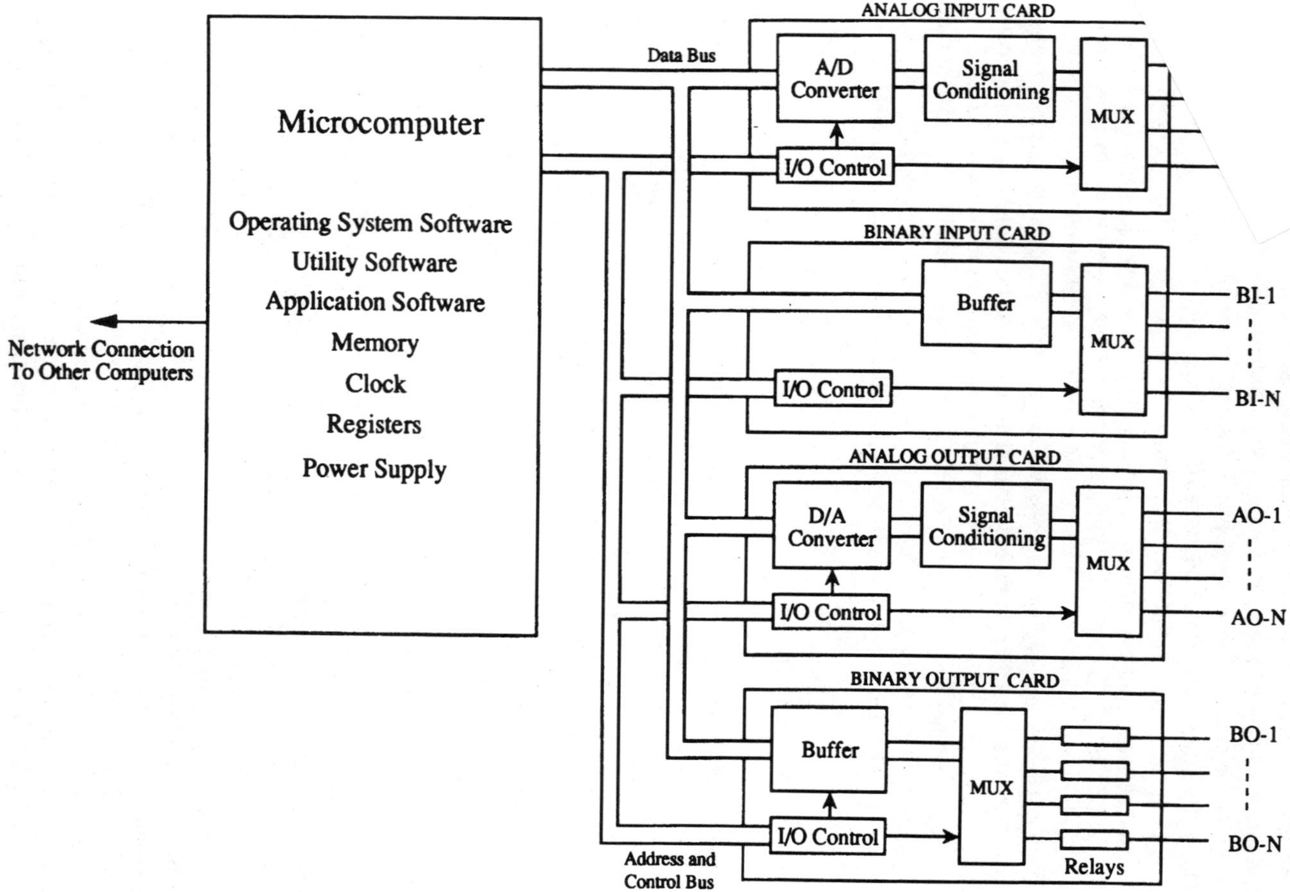

FIGURE 3.3. A block diagram of a typical microcomputer used for DDC.

feedback points and values for the various tuning constants. The actual calculation is then carried out at run time by the preprogrammed code. If some other type of control algorithm is desired, these can usually be implemented by using built-in math and logic routines.

Another distinguishing characteristic of general-purpose DDCs is that they tend to be *modular*. This means that I/O capabilities can be added as required by the job requirements in whatever types and numbers are needed, usually by simply adding another printed circuit card.

Other types of general-purpose computers can be used for DDC. I have a friend who has been controlling his home heating system for years using an old Apple I computer outfitted with a couple of special-purpose I/O boards to interface to temperature sensors and a few binary outputs.

Some plant engineers have made good use of programmable logic controllers as mentioned in Section 2.4.1. It would also be possible, at least in theory, to use industrial-grade process control DDC systems to control HVAC. Like their HVAC counterparts, these devices also provide libraries of functions that can be invoked to implement arbitrarily complex control strategies. But typical industrial DDC libraries would probably *not* provide routines for psychrometric applications such as humidity control or other common HVAC control functions such as optimal start/stop or load shedding, though such capabilities could certainly be programmed.

The main drawback to general-purpose industrial controllers is simply their higher cost relative to commercial controllers. The higher price tag derives from features that may well be worth the extra money if your manufacturing process is at stake. For example, industrial controllers often support redundant communication trunks and redundant power supplies. Systems can be configured with standby processors so that if the primary computer fails, the secondary computer can pick up the former's functions without the slightest glitch. Industrial controllers are also built to much more rigorous standards with respect to electromagnetic and radio frequency interference. If a card-cage-mounted industrial board fails, you can usually plug in a replacement without even having to turn off the power, an action that would be almost guaranteed to cause serious problems with most commercial-grade DDC systems.

3.5.2 Application Specific or Unitary Controllers

Developed in response to the desire for the greater distribution of computing intelligence, smaller DDC devices now exist that are preprogrammed to control specific pieces of HVAC equipment. Typical applications include variable air volume terminal boxes, heat pumps, fan coil units, and packaged air conditioners. These controllers are characterized by the fact that their "personality" is almost entirely determined by their firmware, which is specifically tailored, by the manufacturer, for a particular application. Another hallmark of unitary controllers is their limited I/O capability and relatively low cost.

Consider the case of VAV box controllers. This is an ideal application of unitary controllers because there are so many variations on the same basic theme, each of which can be easily accommodated by relatively minor changes to the firmware programming. There are VAV box algorithms for cooling only, cooling with electric or hot water reheat, cooling with parallel or series fans with electric or hot water reheat, single-duct systems with chilled water valve, dual duct with or without mixing, and so on.

In order to perform the appropriate VAV control function, only a small number of I/O points needs to be provided. These would typically include inputs such as zone temperature, heating and cooling duct temperatures, heating and cooling duct flows, and cooling and reheat valve positions. Outputs would be provided for the box dampers, the valves, and, possibly, the electric reheat circuits. Obviously, the particular inputs and outputs in any given case are dictated by the application. A major difference between general-purpose controllers and unitaries, however, is that once the specific application is determined, the I/O terminal assignments are fixed by the firmware. For example, analog input 1 *must* be connected to the zone temperature sensor; analog output 1 *must* go to the reheat valve, etc.

Some unitary controllers also have operating modes that permit a looser form of application. For example, the same basic controller used for VAV control may also have a decoupled or multiplexer mode wherein the specific purpose of its limited number of analog and binary points is intentionally left unspecified. In this case, these points can be read or written by a higher-level controller just as if they were attached to a passive, dumb multiplexer. This could be useful if what you need to do is simply monitor or control a few random points somewhere.

Finally, some unitary controllers allow you to perform generic closed-loop control for a small number of loops just by specifying the loop tuning parameters. The purpose of the loop can be anything at all, the only constraint being the input/output interface requirements. Thus a unitary could be used to control reheat valves in two or three rooms, two or three electric perimeter heating strips, a couple of domestic hot water heat exchangers, or some other simple application.

3.5.3 "Smart" Sensors and Actuators

The least programmable and, at the present time, least common DDC device is the smart sensor or actuator. The "smartness," limited though it be, is attributed to the fact that the sensor or actuator is packaged with a full-fledged microprocessor or microcontroller. The main purpose of the computer, which may be a single-chip device, is to provide A/D or D/A conversion, in the case of analog points, and a digital communications capability so that the state of the sensor or actuator may be relayed to a higher-level, more sophisticated computer.

The driving force behind smart sensors and actuators is wiring convenience. Instead of having to run separate wires to each sensor and actuator from its associated controller, it may only be necessary to run a common communication cable and power connection to each field device in a daisy-chain or bus topology. The use of a bus, however, introduces a constraint that needs to be kept in mind. Since not all the devices can use the medium at the same time, sampling rate or throughput may be an issue in the case where a smart sensor and smart actuator participate in a feedback control loop administered by a higher-level machine.

The future of the smart sensor concept is intimately linked to data communications. In order for the concept to be viable, the networking technology and the communication protocol or message format to be used would have to be widely agreed upon. Fortunately, progress is being made in the industrial process control industry, an industry whose technology usually leads that of the commercial HVAC controls industry by a few years. The mission of the ISA's SP50 committee is to develop an industrial "fieldbus" standard that would ultimately provide a digital replacement for the ISA-developed 4- to 20- mA analog signaling standard referenced in the last chapter. Although ultimately envisaging the use of wire, fiber optic cable, or radio, the current proposal discusses, among several cabling alternatives, the use of a single twisted wire pair which may, at the user's discretion, combine the power supply and communication function by simultaneously carrying DC voltage and AC data. The ISA's task is complicated by having to consider the issue of intrinsic safety, i.e., making sure that the electrical signals on the wire are capable of passing through potentially explosive atmospheres without setting off any detonations.

Since all analog signal processing is done in the immediate vicinity of the sensor or output device, it should be possible to completely eliminate the kind of errors introduced by long cable runs, including the pick up of electromagnetic or radio frequency interference. Communications protocols with proper error checking should result in the digital information being conveyed with the clarity of a high-fidelity compact disk!

BIBLIOGRAPHY

ANSI X3.4 - 1977, *American National Standard Code for Information Interchange*, New York: ANSI, 1977.

Evans, C., *The Making of the Micro—A History of the Computer*, New York: Van Nostrand, 1981.

Hall, D., *Microprocessors and Interfacing*, New York: McGraw-Hill, 1986.

Liu, Y., *The M68000 Microprocessor Family: Fundamentals of Assembly Language Programming and Interface Design*, Englewood Cliffs, NJ: Prentice-Hall, 1991.

Miller, M., *The 6800 Microprocessor Family: Architecture, Programming, and Applications*, 2nd ed., New York: Macmillan, 1992.

Moreau, R., *The Computer Comes of Age*, Cambridge, MA: MIT Press, 1984.

Subbarao, W., *16/32-Bit Microprocessors: 6800/68010/68020 Software, Hardware, and Design Applications,* New York: Macmillan, 1991.

See also the excellent documentation sets from the chip manufacturers such as Intel and Motorola.

Chapter 4

COOLING CONTROL
SETPOINT AV
AI
AI BI
AO
OUTSIDE AIR TEMP
SETPOINT OCCUPIED
AI
AI
HIGH SELECT
DISCRIMINATOR
ENABLE COOLING

DDC Software

SUMMARY

Software tells the computer hardware what to do. As with the hardware, there are many possible software architectures. Most computers use a hierarchy of programs, or *modules.*

At the lowest level is the *operating system.* This program bridges the gap between hardware-specific functions, such as accessing a particular memory location or outputting a character on a communication trunk, and the application functions which rely on these capabilities and possess a knowledge of the *meaning* of the information being accessed or communicated.

An intermediate level may consist of modules which carry out functions that are intended to be shared among various applications. Such a library of modules might handle communications functions, database management functions, operator-machine interface functions, or specific control algorithms. Their principal characteristic is that they may be invoked, or "called," from various application programs while otherwise being dormant.

At the highest level are the application programs which actually carry out the metering, monitoring, control, or operator interface functions. These programs usually depend on one or more databases containing information about the connected HVAC equipment and the specific functions to be executed.

How, and by whom, each of the software components is generated is largely determined by the computer architecture.

4.1 SOFTWARE CONCEPTS

One of the greatest challenges associated with successfully applying computers to building system monitoring and control is the selection or development of appropriate software. This applies equally to software running in the DDC field panels or on operator workstations, even though the functionality of the software is usually significantly different. In the case of the DDC software, the difficulty stems from defining the control problems to be solved and the subsequent translation of the proposed solutions into working, error-free programs. In the case of the workstation software, the problem is to define the monitoring and supervisory functions desired—and the types of capabilities that are to be provided for the system operators and managers.

In order to understand the different kinds of software found in typical DDC installations, it is useful to arrange the different capabilities in some sort of meaningful way. Several such arrangements are possible. For example, we could classify the software by its location in the system, i.e., where it is to be executed. This would lead to the idea of *workstation software* and *field panel software.*

Another classification could be by function. One such categorization could be *operating system software, utility software,* and *application software.* See Figs. 4.1 and 4.2.

We will take a combined view, looking first at the common aspects of each functional category and then at the differences between the software in each category as it is found in actual workstations and field panels.

4.2 OPERATING SYSTEM SOFTWARE

Every computer has some form of operating system (OS) software to perform the rudimentary computing functions required for the machine to work. These include moving data from one piece of hardware to another, e.g., from a program to a display device or to a communications line; managing memory; servicing interrupts produced by events that occur asynchronously, i.e., without any particular time reference to other events; and, generally, regulating the access to, and use of, the computer's resources including the sequencing and execution of other, more general computer programs.

The simplest OSs support the running of only one program at any given time. Others permit multiple programs consisting of multiple distinct tasks operating in a sort of time-sharing arrangement called *multitasking*. Multitasking allows the computer to appear to the user to be doing more than one thing concurrently, for example, printing a report, accessing a remote computer, and displaying a graphic on a CRT, all at the "same" time.

An operating system determines the character and personality of a computer and how it relates to the rest of the world, including human operators. Since the same computer can run with different OSs, its personality can indeed vary radically! The proof of this can be seen in the fact that many DDCs use the same microprocessors

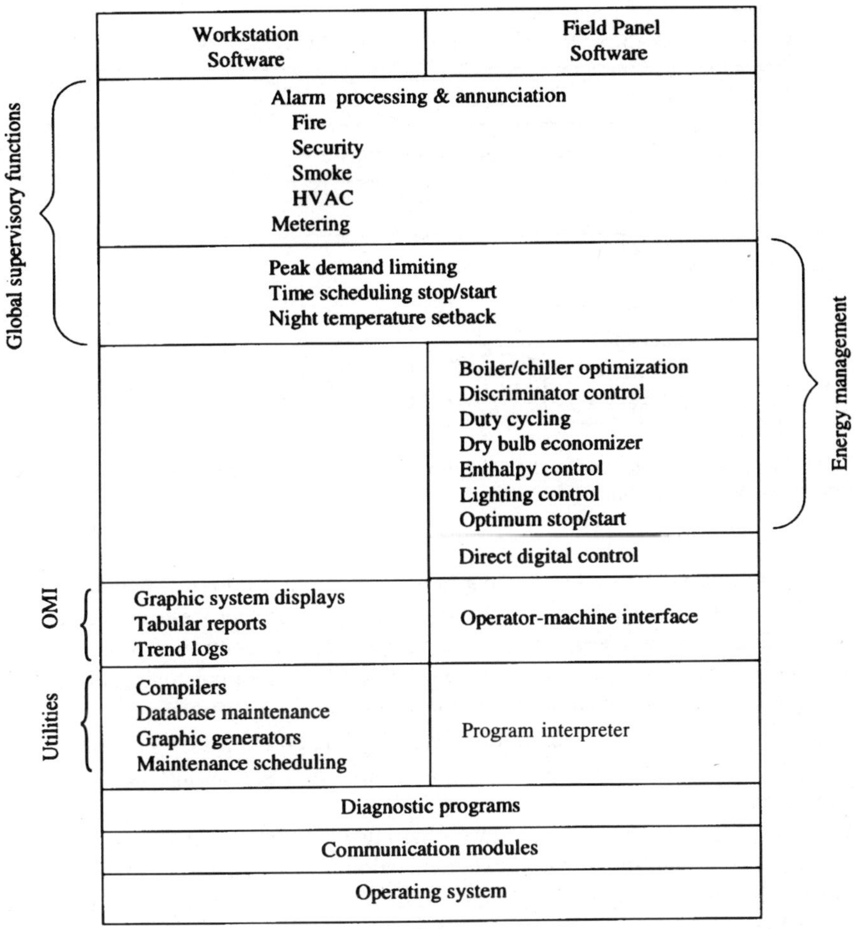

FIGURE 4.1. Two ways of classifying software: by location and functionality. There are, however, often significant differences between host and field panel software in the same general category.

that are found in general-purpose desktop computers—but otherwise share few, if any, operational similarities. Other examples would be PCs running standard PC-DOS one minute and Microsoft Windows the next, or a VAX alternately running VMS then Unix!

Because the OS provides the interface between the computer hardware and other software, an OS is *always* processor-specific, with its routines written in the instruction set of the processor being employed. This is because no two processors have exactly the same instruction set or instruction format. For a typical microprocesser, these "machine" instructions provide capabilities for (1) data transfer, moving data or instructions from one device to another; (2) arithmetic operations, including

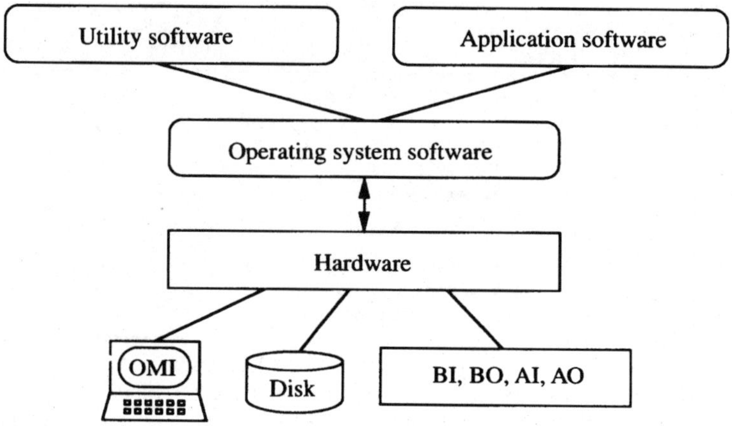

FIGURE 4.2. Both workstations and DDCs make use of three classes of software: operating system, utility, and application.

special operations such as incrementing or decrementing; (3) logical operations, such as performing Boolean ANDing, ORing, inverting, and then testing the resulting values against zero, nonzero, positive, negative, etc.; (4) control transfer from one group of instructions to another by calling subroutines, looping repeatedly through the same instructions until some condition is met, or jumping to some other instruction either conditionally or unconditionally; and, finally, (5) processor control functions, such as waiting for events, clearing interrupts, and scheduling the next instruction to be executed.

While the OS interface to the hardware has always been, and probably always will be, processor-specific, many people have questioned whether or not the interface to higher-level software could be standardized. At present, a program written in a high-level language such as C or Pascal is translated into low-level machine instructions and/or calls to OS routines to carry out the specific functions required by the high-level program by a special program called a *compiler.* If the interface between the high-level languages and operating systems could be standardized by developing common, platform-independent definitions of the OS routines needed to support the compilers of the most common languages, it would be possible to develop compilers that would work with any OS, thus ushering in a new era of program portability. Programs written on one computer could run without modification on any other!

Toward this goal, IEEE Standards Committee 1003 has produced Standard 1003.1-1988, Portable Operating System Interface for Computer Environments (POSIX). Although the initial POSIX specification focuses primarily on the C language running under the Unix operating system, additional work is planned to broaden the standard's scope to apply also to other languages and OSs.

Ultimately, this type of standardization could have a significant impact on the DDC industry by facilitating the development of portable software, particularly for workstations used with field panels from several vendors. Other standards, relating to *windowing,* i.e., displaying several separate terminal sessions on a single screen, and data communications, will also be important in the quest for a more open, standardized computing environment for DDC in the future.

4.2.1 Workstation Operating Systems

Originally, operator workstations were, without exception, simply terminals connected to a central computer. The early computers were often DEC PDP-11's, IBM System/7's, Series/1's, or other computers which had been developed for laboratory or other real-time, process control applications. The earliest "terminals" were actually KSR (keyboard send and receive) teletype machines, clunking along at 5 or 10 characters per second on paper rolls that looked more like Dead Sea scrolls than anything else! Later, of course, the screen-type terminal took over, complete with dynamically updating displays and color. Today, the most prevalent type of workstation is based on a desktop computer, often an IBM 386 or 486, PS/2, or possibly a Macintosh.

All of these workstations, past and present, share one thing in common: the operating system of the underlying computer is standard, off-the-shelf software, thus allowing the control system vendor to concentrate his efforts on building the operator-machine interface software on a known, stable foundation. Examples of these standard OSs are RSX-11M for the DEC PDP-11 computer family; VMS (Virtual Memory System) for the DEC VAX or MicroVAX series; MS-DOS (Microsoft Disk Operating System), RTOS (Real-Time Operating System), OS/2, or Windows for IBM PC's; and the Apple OS for Macintoshes. Most recently, several vendors are supplying systems that use some form of the Unix operating system developed by AT&T. (Unix itself is the target of extensive standardization efforts.)

While a detailed comparison of the attributes of these OSs is outside the scope of this book, two aspects are worth mentioning. First, they all permit operations with disk storage devices. Second, most of them are multitasking OSs and some also support multiple concurrent users.

Another distinguishing attribute is that these standard OSs are *interactive,* responding to commands typed on the operator's keyboard as they are entered. Fortunately, most commercial operator-machine interface (OMI) packages shield the operator from having to know much about the underlying OS by invoking the OMI software at system startup through, for example, an AUTOEXEC.BAT file on an MS-DOS platform.

4.2.2 DDC Field Panel Operating Systems

In sharp contrast to the standardized OSs typically found in workstations, the OSs built into DDCs themselves are totally nonstandard. Although usually based on common microprocessors such as the Intel 8086, 8088, 80186, 80286, 80386, Zilog Z-80, or Motorola MC6800 or 68000 family, the specific functions of each vendor's DDC require an operating system tailored to the individual product.

In many respects, field panel OSs are much simpler than their workstation counterparts. For one thing, field panels generally have no disk drives to worry about (at least not yet!), and their OMI functions are usually primitive compared to workstations. Second, the OS has to be able to fit in ROM. Nonetheless, the OS has to be able to manage data acquisition and control functions, monitoring and control program execution, and, possibly, data communications with other controllers and/or operator workstations. The OS must also be capable of performing diagnostic and self-test functions, either automatically or on the command of a human operator.

4.3 UTILITY SOFTWARE

Although most DDC software runs continuously and automatically, certain types of operations are only performed occasionally and then generally under operator supervision. Such routines are often called *utilities.* Many utilities, particularly in the case of workstations, are general-purpose housekeeping routines and have nothing to do with the workstation's role as a DDC OMI device. Other routines are specifically aimed at providing some DDC testing or maintenance function.

Where the operating system software leaves off and utility software begins is not always clear. Consider, for example, the operator commands associated with MS-DOS. If you look into the operation of DOS, you will find that the commands are divided into two groups: internal and external. The *internal* DOS commands, like DIR to display a list of disk files, COPY to copy files from one disk to another, or MKDIR to create a new directory, are loaded when the PC is started up and are always memory-resident. In contrast, the "external" commands, the category into which most DOS commands fall, reside as miniprograms on the system disk and are only loaded when invoked by the operator.

4.3.1 Workstation Utility Software

The most common utility routines associated with workstations are used for basic system setup and maintenance. Many of these are used to perform disk-related operations. Examples are routines to format disks, both hard and floppy; search for files with particular names or file types; search through disk files to find a particular string of characters; compress disks to remove unused space between files; compare different versions of files to see what may have changed; rename or delete files; and create or delete file directories.

Other utilities are used to prepare programs written in a particular programming language for execution or to help discover the origin of programming errors. These utilities include assemblers, compilers, translators, linkers, loaders, and debuggers. These terms are defined later in this chapter or in the Glossary.

Routines for the performance of troubleshooting and diagnostic functions are also often available as utilities. Some of these run automatically when the workstation is powered up and are part of the power on self-test sequence which resides in ROM but can also be run at other times on operator command.

The more interesting utilities are those specifically used to manage a network of

DDCs. A common requirement is that of uploading or downloading new programs or databases between workstations and DDC field devices. Generally "down" is from the workstation to the field and "up" is the reverse. The theory of "which way is up" has something to do with the greater intelligence always being on top! When the two computers involved are intellectual peers, however, most people are happy to talk simply about file transfer, which is what up- and downloading are really about. Apart from loading new or updated software to a DDC, these utilities are a fundamental way of backing up or saving individual DDC programs in a safe place, a subject discussed further in Section 4.7.

Other utility programs are designed specifically to test data communication channels and equipment or the responsiveness of a particular field panel. We will discuss more about the former in the chapter on communications. Programs of the latter type are available to loop repeatedly while reading a particular input point, to try to cause intermittent errors to show up; to return data in raw binary format to test whether an erroneous reading is in the hardware or the software that converts the raw reading to engineering units; and programs to send commands to the field panel that may not be implemented in higher level OMI software.

Other programs may allow *virtual terminal* access over the network. These are programs that send messages over the communication line to establish a "connection" to the remote DDC as if an operator's terminal were physically plugged into the controller in the mechanical room. Using this kind of program, certain configuration operations may be carried out that may not be available to an unprivileged operator or through the normal OMI software.

Finally, there are routines to perform clock synchronization and resetting, a capability that comes in handy in those areas that observe the time changes that occur in connection with daylight savings time.

4.3.2 DDC Field Panel Utility Software

Compared to the kinds of utilities found in workstations, the capabilities of most DDC field devices are naturally quite limited. The common utilities are diagnostic. Most all DDCs go through self-test routines at system power up, with the results displayed on one or more light-emitting diodes (LEDs) or available through a local alphanumeric readout or remote display terminal. Other systems continually perform diagnostics with similar displays. Still others allow an operator to run diagnostics in the field at the push of a button.

Built-in utilities to facilitate configuration are also common. These may include complete facilities to display and edit the operational DDC programs and related databases from a portable general-purpose terminal, or simply allow access to a technician equipped with an intelligent, vendor-proprietary handheld programmer. Many recent systems allow a technician to connect a terminal or laptop computer to any field panel and gain access to any other device on the communications trunk. This requires, of course, the same type of virtual terminal capability mentioned above.

Lastly, most DDC field devices allow an operator to override both the connected actuators and the individual software control processes at the field device. This capability can be extremely useful both during initial commissioning and later in the event of problems. During commissioning, the override capability allows a technician to manually exercise each output device through its entire range or set it to some intermediate position to check its effect on other system functions. If, later on, a sensor in a control loop should fail causing an associated valve or damper to go fully open or closed, the override capability would allow a technician to temporarily regain some semblance of control until the sensor could be repaired.

4.4 APPLICATION SOFTWARE

Computer programs that perform "useful work" are called *application programs*. In the realm of direct digital control, these are the data acquisition, metering, monitoring, control, energy management and operator-machine interface programs. Since a basic premise of this book is that you already have a reasonably good idea of how you might want to control any particular piece of building equipment, and only need to find out how to do it with it DDC, I'm not going to describe many specific control strategies in detail. Still, if you are a newcomer to building controls, a brief description of some of the types of application programs that are now available will illustrate the spectrum of capabilities. Whether the use of a particular program is appropriate or feasible naturally depends on the circumstances of the job at hand and the system architecture. Applications that require cooperation among many different controllers naturally require a communication network that stand-alone applications do not!

4.4.1 Energy Management Software

1. *Start/Stop Control.* The most basic, and effective, energy saving technique is to turn off energy-consuming systems when they are not needed! Unfortunately, total shutdown is frequently not possible, thus presenting a real challenge to the designer who must look for alternatives. Some of the techniques below can be useful, but we need to recognize up front that they are only capable of delivering a small fraction of the savings achievable with total system shutdown.

2. *Duty Cycling.* Duty cycling is the shutting off of HVAC systems, usually air handlers, for some fraction of their normal run time, for instance, 10 min out of each hour. It gained popularity in the early days of DDC because it was a function that could *only* be performed by computer. For this reason it was a big selling point! While differing opinions exist on the long-term effects of cycling on starters, bearings, fan belts, etc., there is little doubt that the effect of cycling on building occupants can lead to annoyance and complaints. Other techniques which produce comparable savings, such as reducing fan speeds through resheaving or the use of variable-speed drives, have therefore greatly outdistanced duty cycling in popularity.

3. *Load Shedding.* This technique is used in conjunction with hardware and software for sensing electrical demand to limit power demand peaks which, depend-

ing on the specific electric utility rate structure in effect, can have a disproportionate effect on utility costs. The higher the proportion of total load subject to computer control, the greater the potential for being able to effectively control peak demand. The only problem with this strategy, apart from the potential for annoyance mentioned above, occurs if the bulk of the electrical load is *not* sheddable, consisting, for example, of lights, elevators, coffee pots, copy machines, etc. In this case, load shedding will be essentially useless.

4. *Temperature Setback.* This technique involves resetting a heating schedule based upon occupancy and space temperature. Using hot water radiation heating as an example, the heating schedule, i.e., the inverse relationship between outside air temperature and hot water supply setpoint, would be determined by an outside air sensor during occupied times while an indoor sensor and a lower temperature schedule would be computer-selected during unoccupied times.

5. *Dry Bulb Economizer or Enthalpy Control.* The basis of outdoor air economizer cycles in systems which mix the outdoor and return air is to select the airstream which is either coolest in terms of temperature (dry bulb economizer) or in terms of total heat content (enthalpy economizer) to provide for the cooling requirements of a space. Since enthalpy control requires the measurement of relative humidity or dewpoint in addition to dry bulb temperature, it is more expensive to implement and, except in exceptionally humid climates, usually provides little additional saving over the simpler dry bulb technique.

6. *Discriminator Control.* Another technique that lends itself to computerization is to reset an air handler's hot and cold deck or air discharge temperature by sensing the heating and cooling demands from as many zones as possible and determining how much heating or cooling is required to satisfy the zones furthest from setpoint. The hot deck temperature can then be *lowered* to just satisfy the zone calling for the greatest amount of heating, while the cold deck temperature can be *raised* to just satisfy the zone calling for the greatest cooling. A DDC can even be programmed to throw out a bad sensor reading, a problem that can render a pneumatic discriminator worse than useless.

7. *Variable Air Volume (VAV).* The concept of VAV is to condition only the minimum amount of air required for a given space, thus minimizing the amount of heating or cooling energy used. VAV is fundamentally a cooling technique, with the supply air maintained at a temperature sufficiently low to satisfy the zone calling for greatest cooling with its associated VAV box fully open. Because of this, it is most often used in conjunction with other heating sources such as perimeter radiators. In a typical implementation, a zone temperature sensor produces a signal that is used by the DDC to adjust the VAV box's air flow rate by means of a motorized damper. If the temperature in the space falls below the setpoint temperature, the zone damper will modulate down to minimum air volume. If the space temperature falls further, the same temperature signal can cause the DDC to actuate a valve on an optional reheat coil to prevent overcooling of the space.

Two types of VAV are common. The first is pressure-*dependent* VAV. In this case, the VAV box damper is controlled directly by the space temperature sensor, and the volume of air delivered to the space varies directly as the static pressure in the supply duct varies.

In pressure-*independent* VAV, the volume of air supplied to the space is maintained independent of variations in the static pressure in the supply duct. This is accomplished by installing a flow sensor in the VAV box and calculating the volume passing through the box based on a knowledge of the duct area and damper position. The temperature sensor in this case is used to adjust the volumetric flow rate rather than the damper itself.

In both cases, the less flow required, the less air needs to be provided by the central air handler, which can then be throttled back, either with a variable-speed drive or by using variable inlet vanes, resulting in substantial energy savings.

VAV control is an excellent candidate for application-specific, preprogrammed DDCs for several reasons. First, several sensor-actuator control loops are involved, but they all are located in the VAV box. Second, the control of the primary air handler can be based on measurements of static pressure taken in the duct at the VAV box. With the VAV box controllers on a communications network, the pressure and flow information can be immediately shared with the controller of the central system, an efficient arrangement. Finally, VAV boxes are commodity items and the DDCs can be mass produced and installed on the boxes at the factory.

8. *Variable-Speed Pumping*. DDC can be used to control variable-speed drives not only for air handlers but also for pumping applications. A common application is to use DDC to maintain a constant differential pressure across an hydronic system, e.g., a chilled water system.

9. *Optimal Stop/Start*. In those cases where it is possible to turn off a heating or cooling system during occupied periods, the question naturally arises: when is the optimum time to turn the system back on, both in terms of energy and comfort? A number of algorithms have been developed. Some involve look-up tables relating outside air temperature, building mass temperature, and time-to-occupancy. Others rely on measuring the rate of heating or cooling following shutdown and producing a prediction based on curve fitting techniques. While there are still differences of opinion as to the best strategy to use, almost all of them can be readily carried out with even a fairly simple DDC.

10. *Boiler and Chiller Optimization*. In addition to the sort of energy management functions listed above, there are also programs dedicated to the "optimization" of specific devices such as boilers and chillers. The programs are of two basic types: those that deal with improving the *efficiency* of the specific device and those that concentrate on improving the *system* efficiency by providing a better match between the device and the load. In the latter category are programs that reset discharge temperatures based on feedback from selected zones or that select the optimum configuration of boilers or chillers based on a knowledge of their partial load characteristics, in those cases where several boilers or chillers are involved. This technique is usually referred to as *optimal dispatch*.

4.4.2 Metering and Monitoring Software

Because of its basic data acquisition, storage, and communication capabilities, DDC equipment can be used for much more than basic closed-loop control or energy management. In fact, a DDC device with the appropriate number of binary and analog inputs can be a reasonable choice for data acquisition even if its control capabilities are never used! Metering and monitoring are two such common applications.

1. *Metering*. The data acquistion capability of DDCs can be used to meter virtually any utility to corroborate, supplement, or even replace conventional metering. Chilled water, steam, and electricity are three of the most commonly metered utilities. Measuring fluid energy flows is an application particularly suited to the use of computers since a calculation involving several physical measurements is required. In the case of chilled water, for example, the product of water flow and the difference between the supply and return temperatures is needed. For steam, the flow rate needs to be compensated for effects of temperature and pressure. In the case of electricity, a computer can count the pulses generated by standard watt-hour meters to verify the electric company's bills.

Since most DDCs store data in RAM, their long-term storage capacity tends to be quite limited. The most common programming challenge is therefore to figure out how to read the collected data from the field often enough that no data is lost while still taking advantage of the DDC's ability to continue storing even when the central computer or workstation is off-line.

2. *Monitoring*. DDCs often earn their keep by providing advanced warning of impending problems or by providing remote diagnostic capability of trouble that has already occurred. Since sensors and transmitters are available to monitor virtually any system parameter, the challenge to the designer is to select only those points that are most relevant and significant. If the monitoring is intended to generate "alarms" in the event of off-normal conditions, the software may be written to provide a hierarchy of alarm severity. Conditions that are considered worthy of being reported but which require no immediate operator response may simply produce a printout on a hard-copy device. More severe alarms may activate a visual and/or aural signal, print out the nature of the problem, and possibly even tell the operator what action to take. The most severe alarms could be programmed to do all of the above as well as to call someone via an automatic dialer.

One question that arises is where should the alarm logic and processing be done? In the early days, this was not an issue since the data acquisition equipment was not programmable and all processing was done by the central computer. Today, almost any DDC can be configured to do alarm processing and indicate alarm occurrences to a remote operator workstation.

There is a difficulty, however. When all the logic resides in the field, how can an operator easily determine exactly what has gone wrong? Most systems are set up so that an alarm is signaled from the field to the workstation with only a couple of bytes of data on the communication line. It is then the job of the OMI software to generate any alarm messages that the operator will actually see. While a message of the form "BUILDING X: ALARM 10" conveys some information, it would obviously be better if the message read "BUILDING X: AIR HANDLER 2 DISCHARGE TEMPERATURE IS 49.9. SETPOINT IS 55.0. THERE MAY BE A PROBLEM WITH THE PREHEAT COIL!"

The only impediment to having operator software (and operators!) do a much better job of alarm analysis than is today's norm is the fact that the relevant information is usually fragmented. There are the files containing the configuration information for the DDC itself, including the specific logic which has produced the alarm under investigation. There are probably files containing all of the original system design specifications including the intended sequences of operation and possibly even computer-aided drawings of the system schematics. There may be history files that could be used to indicate if a particular alarm has occurred before and, if so, what was done the last time to correct it.

Fortunately, help may be on the way in form of *expert system* programs that can potentially help correlate these diverse sources of information and thus aid in the diagnostic process. We will take up this subject in a bit more detail in the chapter on OMI.

4.4.3 ADDITIONAL APPLICATION SOFTWARE

In addition to the real-time applications already described, DDC computers and workstations can be employed in a wide variety of facilities management functions.

1. *Fire, Security, and Intelligent Buildings*. Since fire and security functions are often computerized, it is possible to conceive of a totally integrated facilities management system combining fire, security, local loop control, metering, monitoring, and energy management functions (and possibly other services such as digital telephone, FAX, word processing, video conferencing, etc.). Such integrated systems, it is argued, could be the stepping stones to a whole new class of "intelligent buildings."

Nonetheless, considerable controversy surrounds the integration concept. With respect to fire and security, advocates point to the economies associated with the maintenance of only one computer as compared with two or more. Antagonists contend that by the time National Fire Protection Association (NFPA) code requirements concerning failure mode operations are met, little additional benefit can be gained over systems with multiple, dedicated CPUs. There is also the practical matter of jurisdiction. It is often the case that fire, security, and other life and public safety issues are the responsibility of different people than those responsible for temperature control and building HVAC.

In the case of intelligent buildings, few would question the wisdom of providing a cable plant that would allow the flexible interconnection of workstations, computers, printers, and similar office automation equipment throughout the building. In fact, considerable work has been done by the Electronic Industries Association Engineering Committee TR-41.8 in the United States and the Canadian Standards Association Committee T530 to develop commercial building wiring standards. These standards, EIA/TIA-568-91 and EIA/TIA-569-90, specify "design and construction practices within and between (primarily commercial) buildings which are in support of telecommunications media and equipment."

Many question, however, whether confidential corporate records should reside on a shared computer or that several companies should share a common telephone system. Although the term "intelligent building" began as a marketing slogan, it is clear that there is compelling logic to standardize at least some of the telecommunications features of commercial buildings. It will be interesting to see what happens in the years ahead!

2. *Maintenance Scheduling and Inventory Control.* Two useful database management applications are maintenance scheduling and inventory control. Maintenance scheduling is a natural extension of DDC equipment, if so programmed, to accumulate run-time totals for fans and pumps. Based upon a predetermined preventive maintenance schedule, the computer can generate detailed work orders to guide maintenance personnel. Following the conclusion of the work, the parts used can be deducted from the inventory database.

Two other important software topics are communications software and operator-machine interface software. Stand-alone systems may require no external communication capability and a limited OMI. In other cases, the success of an installation may hinge directly on the effectiveness of its networking and OMI software. These topics are covered in chapters 5, 6, and 7.

4.5 CONFIGURING DDCs: OVERVIEW

The process of configuring a DDC to perform useful tasks generally involves two parts: the first is to define a database of information that the controller can refer to in carrying out its task; the second is to provide the specific logic of the task itself. The degree to which the configuration process involves each of these parts is determined by where the DDC falls on the continuum between application-specific, dedicated controllers and multifunction, general-purpose controllers.

In the case of application-specific controllers, the logic has been built into the machine by the manufacturer and generally resides in firmware. An example would be a DDC to control a VAV box. In this case, even the sensor and actuator connections have been preordained. About all the configuration process entails is providing data about the specific VAV box: airflow sensor calibration data, operating and alarm setpoints, and duct parameters.

In the case of general-purpose controllers, all operating parameters and logic are installed in the field. Each input and output point must be defined as well as any logic that exists between them. The beauty of a general-purpose DDC is that you can specify any relationship between the inputs and outputs that suits your application—then readily change it later if the application changes or you come upon a more brilliant, innovative way of performing it!

In most commercial DDCs, the user has a vendor-supplied "library" of relationships to choose from and numerous "functions" that provide specialized capabilities for use in HVAC control. How many, or few, of these capabilities the user ends up calling upon depends on the requirements of the application and his skill and ingenuity. (As with all computer programming, some solutions may be more "elegant" than others!) Finally, the user can always create his own specialized routines to carry out functions not provided by the library.

4.6 CONFIGURATION METHODS

DDCs can be configured either *directly* or *indirectly*. In direct methods, you communicate directly with the DDC through an operator interface such as a keypad and alphanumeric display built into the DDC or through some type of portable operator terminal. The latter may be a general-purpose portable computer terminal (even not-so-portable units have occasionally been pressed into service) or a special vendor-specific handheld unit with special keys and display, designed to expedite the configuring process.

In indirect methods, you generate the necessary configuration data by means of computer programs that run on non-DDC computers, typically desktop systems such as IBM PCs. This process is often called *computer-aided engineering* (CAE). The ultimate purpose of these programs is to collect and format enough data to completely specify the operation of the DDC. The data, once collected and formatted, are stored in files on the CAE computer and may then be downloaded to the DDC at the convenience of the user. The CAE approach provides a built-in backup capability, since the DDC configuration information originates on the CAE machine. The CAE machine also generally contains a program that can print some type of listing or documentation that can be used for subsequent reference or troubleshooting.

Let's now look at several widely used CAE configuration techniques in more detail. Remember, the configuration of real systems is usually performed using some combination of these and that the final tune-up of the DDC is usually performed in the field directly.

4.6.1 Database Entry

The three most common CAE methods of supplying database parameters are (1) text entry using a text editor program; (2) entering data in response to CAE prompts; and, (3) filling in fields on an entry screen, or template. Each of these methods has its pros and cons.

The main advantage of text entry is that many parameters can be typed in on a single line. Another advantage is that text tends to be self-documenting: if you know the entry format or syntax you can *see* directly if the entry is correct. Also, you can usually add comments to annotate the entry further if it isn't self-explanatory. Finally, since the data is in a text file, you can easily run off a hard copy of your work for review purposes.

There are several disadvantages to text entry. The first is that you have to know how to use the text editor program. This isn't much of a problem unless you rarely use a computer in your day-to-day work. Second, you have to know the exact syntax of the database entry statements, i.e., where every comma and semicolon goes and in exactly what order. Finally, the text must usually be subjected to additional processing before it is ready to be loaded into the DDC. This means knowing how to run yet another program such as a translator or compiler.

The main advantage of the prompt-response method is that the CAE program *asks* for the database information it needs to properly define the DDC configuration, thus freeing you from the burden of having to remember particular keywords or formats. Most CAE programs will even suggest the range of appropriate values for your entry or possibly provide a default value that you can use as is, or edit if you find the default inappropriate for your particular application. The CAE program may also be smart enough to tell you if you have entered an incorrect value. In the case of text entry, you usually don't learn of your mistakes until you run the translator program.

The drawbacks to the prompt-response method are minimal. Some of the early CAE programs could "sneak up" on an unsuspecting user, asking questions the user was unprepared to answer, or responding uncharitably to input that was out of range. Documentation of the configuration data you have entered is another issue. This is not a problem as long as the CAE system you are working with has a way of extracting your original input from the configuration file it produces in a form that can be viewed on the screen or sent to a printer.

Of the three database entry techniques we are considering, the entry screen or template method is far and away the most user-friendly. In this case, the CAE program generates a screen display with a space, or "field," reserved for each required information item. You then simply move around the screen, using a mouse or the cursor keys, filling in the appropriate values. When you are finished, you hit the <ENTER> key and the CAE program processes the entire set of data you have entered, then sends it to the output file.

Again, the documentation of the configuration information you have entered is the major concern.

4.6.2 DDC Programming

Much of the satisfaction that you can have in working with DDC comes from the process of developing and installing the control logic that transforms an inherently dumb machine into one that carries out your exact wishes, perhaps even with elegance and flair! Most often, this process still involves the use of procedural, text-based programming languages. This "traditional" approach to computer programming is certain to be replaced in the 1990s by far more user-friendly, nonprocedural techniques such as graphical programming.

The distinction between "procedural" and "nonprocedural" programming is noteworthy and is receiving increasing attention in the computer industry at large. Procedural techniques tell a computer *how* to do something; nonprocedural techniques tell a computer *what* to do. So-called fourth-generation computer languages are nonprocedural. An analogy used by J. Martin (Martin 1985) to describe the differences in these concepts is that of giving instructions to a taxi driver. You could say "Drive down this street for one-half mile, turn left at the light, go another half mile and turn right . . ." or you could say "Take me to the Acme Movie Theatre." In the first instance, your instructions are procedural; you tell the cabbie exactly how to get to your destination. In the second case, you simply tell the cabbie where you want to end up; the procedure for getting you there is up to him. (In some cities, this can be expensive!) Nonprocedural DDC programming would therefore facilitate the development of DDC software by people who know exactly what they want the controller to do but are perhaps less schooled in the details of telling the controller how to do it.

4.6.2.1 DDC Programming Languages The text-based programming languages for DDC can be divided into two classes: "standard" high-level languages, and vendor-specific languages.

"Standard" High-Level Languages The word *standard* is in quotes because the actual ANSI standard forms of the common languages have no facilities for performing the real-world input/output functions that characterize DDC and other process control applications. Standard languages must thus be extended to incorporate these capabilities. In addition, most high-level language implementations used for DDC programming have been augmented by the provision of functions or subroutines that the programmer can invoke to carry out process control or HVAC-related calculations. Examples include routines to carry out averaging, high-low selection of input values, PID loop control, and psychrometric calculations.

Standard high-level DDC language implementations are most often based on BASIC, FORTRAN, and Pascal. BASIC is an acronym for Beginners All-purpose Symbolic Instruction Code. FORTRAN, similarly, stands for FORmula TRANslation. And Pascal is a language named in honor of the French mathematician Blaise Pascal who, in 1690 at the age of 18, invented the first mechanical calculating machine.

Another language, C, is also being used increasingly. C shares many of the high-level characteristics of Pascal, such as the availability of pointers and structured data types, but also allows many low-level operations, such as easy bit manipulation, more commonly found only in assembly languages. Using certain of C's preprocessor directives, such as the ability to define aliases for some of C's more cryptic syntax, it is possible to use C to create new languages with English words (or those of any other language) while retaining the power of the underlying language (Petze and Chaudry 1989).

Examples of the use of extended high-level languages are Landis and Gyr Powers' Powers Process Control Language (PPCL), a FORTRAN derivative, Honeywell's DeltaNet Pascal, and Johnson's JC-BASIC. PPCL is used to program Standalone

Control Units (SCU), the DDC component of the System 600. See Fig. 4.3. DeltaNet Pascal is used to write "custom DDC" programs for the Excel Plus line of DDCs. JC-BASIC is used with, or without, Johnson's graphical programming language for Metasys products.

Vendor-Specific Languages While some vendors have chosen to modify or extend existing languages, others have chosen to construct programming languages from scratch. The language used to program the American Auto-Matrix STAR, though currently unnamed, is a descendant of the Rex Programming Language (RPL) used with its earlier controllers. Andover also uses an unnamed vendor-specific language to program its AC256M Plus master panels and its Local Control Units (LCUs). The LCUs, once programmed, can operate without the presence of a master panel. Johnson Controls uses Custom Application Language 1 (CAL1) with its DSC-8500 controllers. (DSC stands for Digital System Controller.)

4.6.2.2 Translators and Compilers An advantage of the direct configuration methods, where you enter database information and logic directly into the DDC that will carry out the monitoring and control functions, is that you are finished with the process once the information has been entered. In most systems using a host computer to process text-based programming languages, an additional step is usually required. This step involves running a program that converts the text file containing the program into some type of machine-readable binary format which can then be loaded into the actual controller. These programs are usually referred to as *translators* or *compilers.* See Fig. 4.4.

The precise distinction between a translator and a compiler is somewhat arbitrary.

```
100  IF(TIME.GT.8:00.AND.TIME.LE.17:00) THEN
+    ACT(120,130) ELSE DEACT(120,130)

110  IF(TIME.LE.8:00.OR.TIME.GT.17:00) THEN
+    ACT(140,150) ELSE DEACT(140,150)

120  ON(SFAN1)

130  LOOP(128,MATEMP,DMPR1,MASP,4000,
+    400,0,1,5000,113,183,0)

140  OFF(SFAN1)

150  DMPR1 = 1.0
```

FIGURE 4.3. Example of a small text-based routine to control a mixed air damper. Between 8 a.m. and 5 p.m. lines 120 and 130 are activated, turning on supply fan 1 (binary output SFAN1) and placing the dampers (analog output DMPR1) under control of the LOOP statement. The parameters specified in the LOOP statement provide the values needed for the DDC to carry out proportional plus integral plus derivative (PID) control. Before 8 a.m. and after 5 p.m. lines 140 and 150 are activated, turning off the fan and closing the dampers. The language is Landis & Gyr Powers' PPCL.

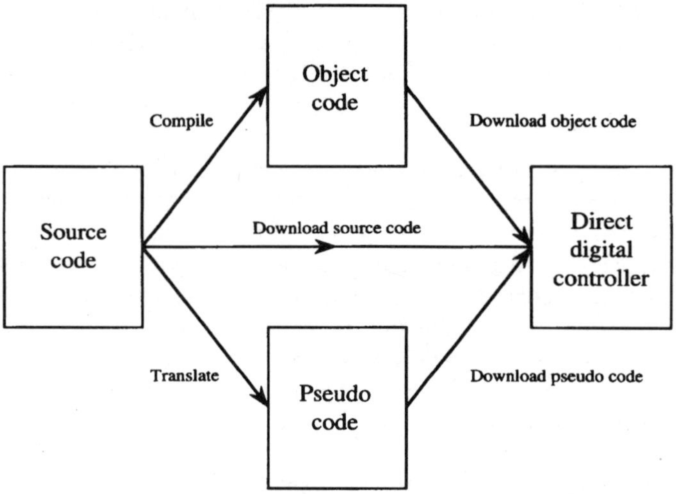

FIGURE 4.4. Depending on the DDC, program source code may be downloaded directly, then, interpreted in real time line-by-line, or first compiled into object code or translated into pseudo code.

A translator is usually a program that converts the input text file into an equivalent, compressed, binary representation. Because the output file represents the source code, but in a binary, nontext format, it is often referred to as *pseudo code.* The pseudo code representation of the configuration data is then interpreted by the operating system of the DDC to carry out the control program. Johnson Controls' CAL1 programs, for example, are processed in just this way.

Compilers, on the other hand, convert the input text file into actual machine instructions of the microprocessor used in the DDC. While pseudo code could, at least in theory, be interpreted by operating systems running on various CPUs, compiled machine instructions can only be executed by the particular processor for which the compiler was written.

In either case, the ultimate purpose of translators and compilers is the same: to convert a human-readable text file into something that a computer can more readily interpret. An additional purpose is to reduce the size of the input file so it takes up less space in a DDC which may, for reasons of economics, have limited memory capacity.

To understand this, remember that text files consist of a binary representation of *each character* in the file, usually consuming one byte per character. Thus it is often possible for a translator or compiler to take a string of characters (assuming the string represents some particular computer *operation*) and reduce it to a numeric code representing the same operation occupying perhaps only a single byte.

The same applies to textual representations of numbers. For example, the three-byte character string '123', representing the decimal number 123, can be represented in one byte as hexadecimal '7B'.

4.6.3 Graphical Programming and Other Techniques

While it is true that today text-based programming continues to prevail, DDC configuring in the future will surely be carried out by nonprocedural means. At least two strong forces will contribute to this evolution. The first will be the ready availability of computing power far in excess of what it available today, at a fraction of the cost. The second will be the universal desire to spend less time worrying about the intricacies of programming and more time solving the problems of building automation and control.

Graphical programming is one nonprocedural technique already available to a limited degree. It uses computer-generated graphical symbols to describe pictorially which control sequences are to be executed. Using "engineering workstation" software, the programmer places boxes on the screen representing the inputs, outputs, and functions to be performed. Each function block the programmer references (by means of its unique function code or icon) is automatically displayed with its title and the appropriate number of input and output line "stubs." The boxes are then connected by drawing lines between the stubs. When the schematic drawing is complete, the computer converts the drawing into an executable format that may then be downloaded to a controller. In addition, the programmer can print a hard copy of the drawing to document the "as-built" control configuration.

Function blocks representing complex operations, such as a PID control loop, are usually associated with numerous parameters that must also be supplied. This is often accomplished by bringing up a template screen, perhaps by double clicking on the function block icon, and filling in the appropriate data. See Fig. 4.5.

Eventually, computers will be smart enough to process natural human language, both written and spoken. "Natural" means language as used colloquially, without special mnemonic terms or syntax. The main advantage of human language is that it would free programmers from the burden of having to learn complex computer language syntax. It would also allow individuals, perhaps with plentiful controls experience but little computer experience, to carry out DDC configuration.

Unfortunately, human languages are notoriously imprecise and ambiguous. We tend to resolve these ambiguities from the "context" of the remark or by the use of "common sense." So far, at least, attempts to instruct computers in either of these techniques have not been particularly successful. One illustrative problem is that the same word can be in one instance a verb and in another instance a noun. Consider the sentence "time flies." Most of us would tend to think of "time" as the noun and "flies" as the verb, simply because someone with a stopwatch, timing flies, is not a commonplace occurrence. To a computer, however, the sentence is totally ambiguous.

The ultimate solution of the problem is probably to combine natural language with

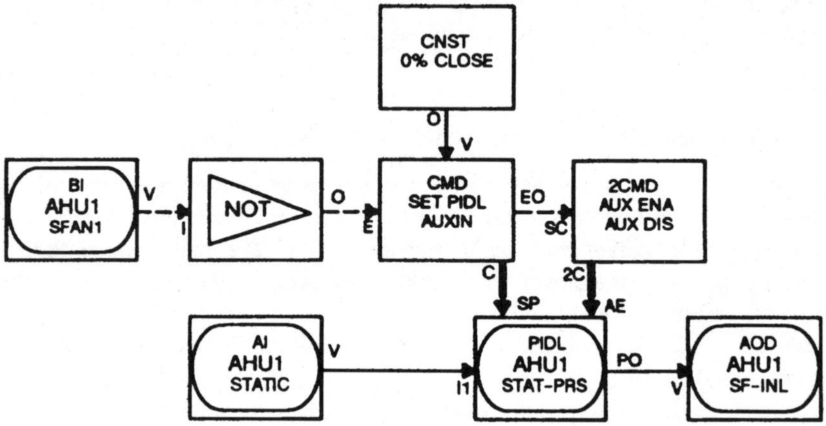

FIGURE 4.5. Example of a Johnson Controls Metasys Graphic Programming Language (GPL) routine to establish a PID loop to control the static pressure of an air handler. The blocks with the racetrack pattern inside are associated with software "objects" each of which is a collection of information relating to a particular function. By positioning the cursor on such a block and double clicking the mouse button, a template can be accessed that allows the programmer to specify the object's parameters.

other techniques such as data entry screens, feedback from the computer to resolve ambiguities, graphical schematic input, and so on.

4.7 BACKUP

Regardless of the configuration methodology, it is of paramount importance to store a copy of the configuration data in a form that can be readily retrieved in the event of hardware problems with the DDC. Such a copy of the configuration data is called a *backup.*

The precise method of performing a backing-up operation depends on how the DDC was configured at the start and what communication facilities, if any, are available. If the DDC was configured directly, the configuration data can generally be dumped to a cassette tape or a floppy disk by means of an appropriate recording device. If remote communications with the DDC exist, it may be possible to send a command to the DDC that will cause it to send a copy of its configuration data over the communication line for storage in a remote computer.

If the DDC was configured indirectly to begin with, the problem is much simpler since a copy of the configuration information already exists on the computer used to develop the configuration file. In this case, it is only necessary to follow normal practice and store a copy of the data in a secure place!

REFERENCES

EIA/TIA-568-91, *Commercial Building Telecommunications Wiring Standard*, Washington: EIA, 1991.

EIA/TIA-569-90, *Commercial Building Standard for Telecommunications Pathways & Spaces*, Washington: EIA, 1990.

IEEE Standard 1003.1-1988, *Portable Operating System Interface for Computer Environments (POSIX)*, New York: IEEE, 1988.

Martin, J., *Fourth-Generation Languages, Volume I, Principles*, Englewood Cliffs, NJ: Prentice-Hall, 1985.

Petze, J. D., and Chaudry, S., An Industry Standard Language for DDC Systems: A Look at the 'C' Language, *ASHRAE Trans.*, Vol. 95, Pt. 1, Paper CH-89-4-3, 1989.

Chapter 5

The Operator-Machine Interface

SUMMARY

The Operator-Machine Interface (OMI) consists of the hardware and software that allow a human operator to communicate with the computer.

OMIs vary greatly in sophistication and capability, depending largely on their location within the distributed architecture. Most support bidirectional dialog. Input devices range from a single switch or push button to full keyboards, light pens, or mice; output devices may be as simple as a light emitting diode (LED) or as refined as a full-color CRT or speech synthesizer.

The OMI of the future will likely take full advantage of the availability of increased computing power by implementing software that allows multiple displays or windows on the same screen, "point-and-click" access to graphic displays and system functions, and the latest developments in artificial intelligence (AI).

In fact, the greatest single factor differentiating today's OMIs from tomorrow's will probably be the increasing application of *knowledge-based engineering* (KBE), one of the branches of AI. KBE techniques can be used to match observations of conditions within an HVAC system with various preestablished rules to allow the computer to draw inferences about the observations. Such inferences could be used to diagnose malfunctions, adaptively fine-tune control loops, and make system adjustments to minimize energy use.

The result of a higher level of computerization could be more efficient, intelligent, and cost-effective operation of the HVAC systems and the transition of operating personnel from the role of system "operator" to that of system "manager."

5.1 FIELD DEVICE OMIs: BUILT-IN

The sophistication of the OMIs provided for direct communication with a field device *at the device,* in contrast to those available using a remote network connection, is highly variable. At one extreme, the field device may display no indication at all that it is functioning although even the least expensive device will usually be equipped with an LED indicating the presence or absence of AC power. If the device is configured for remote access, there will likewise almost invariably be one or two LEDs indicating whether network communications are occurring. One LED typically shows the presence of received data while another shows that the local device is originating a transmission. There may also be LEDs to indicate whether local diagnostic circuitry has detected any problems, whether any outputs are overridden, and whether any predefined alarm conditions have been detected.

On the input side, the simplest devices may have nothing more than a on/off switch and perhaps a momentary contact push-button switch that causes the machine to go through a self-test process.

At the next level of capability are OMIs with one or more seven-segment LED or liquid crystal digits that can display a limited amount of numeric, or alphanumeric, data. Queries or commands are typically entered using a keypad equipped with numerals and one or more special function keys. The functions that can be accessed may be hard-coded into the display firmware or programmable. In the latter case, the function may be associated with a numeric code. Function code 123, for example, might tell the machine to display the present value of a specific physical or logical input or output or allow the operator to set an analog or binary point to a desired value by entering additional keystrokes. Other hard-coded or programmable functions include displaying controller parameters related to time such as the current date, time, and day of week; self-test results; counts of the number of alarms or overridden points; and, possibly, an indication of which points are in fact in alarm or overridden.

At the top end of built-in OMIs are those that provide a CRT-style, 40- or 80-character per line alphanumeric display. Although single-line displays seem to be most common, some DDCs do feature multiline screens. Such displays generally use 7 x 9 or 9 x 12 dot LED or liquid crystal matrices, the latter often back lit for viewing in low-light conditions. The keypad may be a QWERTY, i.e., typewriter style, keyboard or outfitted with special function keys. As with the other built-in OMIs, the functionality varies from one device to the next. If the machine is one of the general-purpose class of controllers and the operating system interprets text-based source code on the fly, it may be possible to display the actual control program line by line and even edit it in the field. While this may be desirable or necessary, particularly during the commissioning phase of an installation, beware of the potential concurrency problems that may arise if the modified program is not uploaded to some safe repository for backup purposes! This same caveat applies to the less comprehensive OMIs, many of which potentially allow an operator to modify some or all of the operating parameters used by the control processes, e.g., analog setpoints, alarm limits, and so on.

5.2 FIELD DEVICE OMIs: PORTABLE

Devices that do not have sophisticated built-in OMIs may still allow a complete range of operator interaction in the field by means of one of several types of *portable* OMIs. The two most common are the general-purpose laptop computer and the special-purpose, vendor-proprietary terminal device.

Laptop computers are popular because of the inherent power they bring to the OMI function. If OMI display software resides in the DDC device, the laptop may only have to function as a "dumb" terminal using some form of terminal emulation software. (This mode of operation is also possible using an *actual* dumb terminal but such terminals are rarely portable.) Alternatively, the laptop may run an OMI program that provides a graphical or menu-driven interface. In this type of operation the connection to the DDC may be via an EIA-232 connection or via a network port. Depending on the network architecture employed, the laptop may run software that makes it appear to the DDC just like any other operator workstation communicating over the network. The connection via a dedicated local connection, on the other hand, may allow the laptop to perform certain types of configuration or diagnostic activities that are only permitted to operators with direct, physical access to the controller.

Special-purpose, vendor-proprietary terminals are often about the size and appearance of a handheld calculator. They typically contain firmware which generates messages to the controller in response to a very few keystrokes. They may have menus built in through which the operator navigates by arrow keys on the keypad. One recently introduced terminal of this genre has no keypad at all but rather a touch-sensitive, sealed membrane "touchpad." The display of this particular device is a 256×128 liquid crystal dot matrix. This allows not only characters to be displayed but also vertical and horizontal lines, which can be used to divide up the screen into windows. In operation, the display software prompts the operator for inputs which are made by pressing spots on the touchpad corresponding to areas of the screen which take on different meanings depending on the function being carried out. The touchpad itself has no markings of any kind, which takes a bit of getting used to!

One of the differentiating factors between different vendors' field OMIs is the degree to which they provide dynamic update. OMIs that don't automatically update their displays are, in my opinion, a true nuisance! What most operators are typically trying to do is to troubleshoot some problem with the HVAC system. To do this you want to be able to display one or more values from the DDC and see if they correspond to the actual state of the HVAC system in question. For example, if an air handler is running, does the DDC panel correctly display its status? If a valve is fully closed, does the DDC know it or does it think the valve is in some other position? If changes are made to the system, you want to be able to see these changes immediately reflected in the OMI display. To be forced to re-display the various points to see if the DDC has detected their change is, as I've suggested, a real pain!

Another differentiating factor between field OMIs is the degree of access security they provide. Some vendors implement virtually no security at the field device, presumably on the theory that anyone who has gained access to the mechanical room and to the DDC itself probably is authorized to manipulate the operation of the controller. Other vendors provide elaborate security schemes which may involve the same type of login procedures, complete with user identifiers and passwords, that are routinely encountered with head-end workstations. To what extent such measures are appropriate or desirable depends on individual circumstances. I am a firm believer in security measures where the risks are great, e.g., at the central computer where an unskilled operator could inadvertently wreak havoc by accidentally deleting files or issuing some ill-advised global command. I am more of a skeptic in the field environment where tradespeople simply want to get in, fix what is wrong, and get out. Anything that adds keystrokes and time to the work is probably not worth it!

5.3 WORKSTATION OMIs: HARDWARE

I have had the good luck to experience about three quarters of the computer revolution first hand. During the last 30 odd years, starting roughly in 1960, computers have gone from the behemoth mainframes where only a handful of people actually came in direct contact with the machines (others communicated via punched cards and paper printouts), to "time-sharing" systems where dumb terminals proliferated and multiuser operating systems like IBM's CMS and DEC's VMS were developed, to today's desktop computer world where machines with far more computing power than the original mainframes are now about the size of a breadbox! With the advances in networking, distributed computing is fast becoming not only possible but the norm. "Distributed" means that information may reside anywhere in the network and be accessed by any machine as if it were local to that machine. These same trends and possibilities have also made their way into the realm of DDC.

The first use of computers for building automation and control involved central minicomputers which performed strictly a supervisory role, monitoring alarm points and perhaps carrying out certain global operations such as starting and stopping entire air handler systems. Local loop control was still pneumatic. The OMI to these earliest systems was a teletype machine, capable of clunking out 5 or 10 characters per second on a paper roll. Later, when the serial, asynchronous data communication interface standard RS-232 was introduced in the 1970s, a single black and white alphanumeric terminal became the OMI of choice. RS-232 (Recommended Standard-232 of the Electronic Industries Association, now officially called EIA-232) in fact precipitated a revolution in terminal devices. Prior to its introduction terminals were strictly proprietary. If you had an IBM computer, you had to go to IBM to buy a terminal!

The first affordable cathode ray tube, or CRT, graphic capability was introduced in the early 1980s. These terminals were based on "stroke writing" rather than on television style raster line displays and exemplified one solution to the two main issues in designing a CRT display: (1) how is the image drawn on the screen and (2) how is it held steady? In the case of stroke writing terminals, lines were drawn at any angle by using the magnetic deflection coils to sweep the electron beam in the desired direction; stroke-written letters were created the same way, by drawing them on the screen. The image was held steady by directing the electron beam onto a phosphorescent screen to which high voltage was applied. The screen could be manually erased by pressing a key which would apply a pulse of reverse polarity high voltage to the screen, discharging the phosphors. The two obvious problems with this approach were that anything drawn on the screen would stay there until it was erased and there was no possibility of color.

Raster technology, of course, revolutionized the CRT industry. In this technology, the electron beam is drawn horizontally across the screen hundreds of times per second, each time at a slightly different vertical displacement, until the entire screen has been illuminated. As the beam sweeps across, its intensity and vertical position within the raster line are modulated to cause differing numbers of electrons to strike phosphors particularly formulated to emit red, green, or blue light. Red, green, and blue are called the additive primary colors. If all are present and equally intense, white is produced. By repeating this process 30 or 60 times a second, a full-color, stable display is produced.

The thing that made early color CRTs expensive was the necessity to "rasterize" the computer image. Consider a line drawn at 45° to the horizontal. The computer would send the terminal a command to draw a line from $(x,y) = (0,0)$ to $(100,100)$. The terminal would then have to figure out which portion of which raster lines to intensify so that the eye would perceive a "line" directed at the proper angle. The reason color TVs cost so much less was that the rasterization of the image was already done at the studio and the number of sweep lines required for an image intended to be viewed from 10 feet away could afford to be far less than for the higher resolution images intended to be viewed from a foot or two away.

Early computer graphic images were still primitive compared to what is available today. The main advances have come about as the result of dramatic improvements in CRT technology and the tremendous drop in the cost of computer memory. These two developments have made possible "bit-mapped" graphics. In bit-mapped graphics, the display image is considered to be a matrix of picture elements, or *pixels,* each of which is identified by its x and y coordinates. Low resolution displays might have 512 rows of 780 pixels while a higher resolution, and cost, display might be 1024 × 1560 or greater. The information about each pixel is stored in one or more bits that comprise the map of the display. If only one bit per pixel is available, (the pixel "depth" is 1), the bit would simply indicate whether the pixel should be illuminated or not. If additional bits are available, color information can be represented. Eight bits per pixel are common, allowing for 256 colors. The screen I am looking at as I type has 72 pixels per inch and a pixel depth of 8.

The trend toward smarter CRT displays and ever smaller computers has resulted in what we call today workstations. Exactly where the dividing line between a workstation and a personal computer is located is a matter of debate. Most would argue that a workstation requires the presence of a networked host computer in order to function. But most workstations are now available with hard disks and can perform tasks exactly like a PC. In any case, the future clearly belongs to OMIs with the power of PCs and the network capabilities of workstations. From here on out, let's refer to the operator's monitor/CRT/PC as simply a workstation.

Today's most common operator workstation is, in fact, an IBM-compatible personal computer. In fact, I know of no current DDC manufacturer (in 1993) who does not provide an OMI based on this platform. The only real problem with the PC-based OMI has been the fact that its DOS operating system has always been oriented toward a single user. This has made it a somewhat inappropriate system for larger installations where many people may want to interact with the DDC field panels, or accumulated historical information, more or less concurrently. As a result, some building automation and control systems, such as the one here at Cornell University, are built around computers like Digital Equipment Corporation's VAX, which feature true multiuser, multitasking operating systems and can be readily networked to PCs, Apple Macintoshes, and a wide variety of other types of workstations.

Just as the color CRT is the most common output device, the general-purpose keyboard remains the most common input device. As time goes on, however, the point-and-click style user interface developed by Xerox and introduced to the common man by Apple, will undoubtedly gain increasing prominence. The reason is simple: it is easy to use! Instead of having to remember arcane commands, the user simply moves the cursor, usually with a mouse, to a menu item corresponding to the desired action. A basic list of such items is always on the display and, once selected, can be "pulled down" or expanded to provide ever more narrowly defined choices of action. Workstations can also be equipped with cursor-movement devices popularized in the world of computer-aided design and drafting such as light pens and trackballs.

5.4 WORKSTATION OMIs: SOFTWARE

As with all computer systems, it is the software that really provides most of the desired capabilities. Let's look at some of these capabilities in more detail.

- *Programming/Configuration.* This is certainly one of the most important workstation functions, and the various methods have already been discussed in the previous chapter. The workstation is also usually the repository for DDC program backups. The backup copies of the programs may reside on a local hard disk or be copied off onto floppy disks. Backup tape systems that make use of 4-mm digital audio tape (DAT) are also starting to come into widespread use. DAT is an incredible technology. One tape, about $2 \times 3 \times \frac{1}{4}$ in., can store over 2 *gigabytes* of information! That's about 2000 floppy disks!

- *Upload/Download.* Along with programming, configuration, and storage capabilities, workstations usually have programs to allow an operator to download program information to DDC field panels or upload the current field panel configuration to the workstation.
- *Virtual Terminal.* Virtual terminal (VT) software allows an operator to communicate with similarly programmed field devices across a network connection as if the workstation were plugged directly into the field panel in the mechanical room. As a result, certain types of privileged configuration or diagnostic operations may be possible that would otherwise require a trip to the field. The intelligence for VT operations usually resides in the field panel, which provides whatever operator interface it desires. The workstation software usually consists of a terminal emulation program and the necessary communication routines to format VT network messages. BACnet provides a VT capability that is described in Chapter 7.

While the software capabilities just presented are important, they all revolve around the configuration and maintenance aspects of the distributed DDC system itself. On a percentage basis, most workstation use is undoubtedly devoted to overseeing the building systems, a good thing indeed since this is how most workstations earn their keep! Here are some of the software capabilities that you can expect to find.

- *Tabular Reports.* Tabular reports take on many forms and have many uses. One of the most common, and important, is the reporting of "alarms." An alarm in this context is any building system event that is significant enough to be called to the attention of a human operator for possible action. An example would be an air handler discharge temperature rising above a certain threshold, perhaps subject to the constraint that the fan had been running for at least 5 min. Another example would be a discrepancy between the desired status of a fan or pump and the status sensed by means of a flow switch.

The determination that such an event had occurred was traditionally performed at the central computer, which maintained the setpoints and algorithms necessary to perform the calculations. With distributed intelligence, however, it is now much more common that the calculations are performed in the field, with only the fact of the alarm's occurrence reported back to an operator workstation. While this generally yields faster response times since alarm calculations can, in effect, be carried out in parallel throughout the network, there is a potential database problem involved. The problem is that the first thing an operator wants to investigate when an alarm is reported is whether or not the alarm is legitimate, i.e., is there truly a problem with the HVAC equipment or is there a problem with the monitoring equipment? In order to make this assessment, the operator needs to know how the device reporting the alarm made *its* determination. To do this, of course, the operator has to have a way of accessing the logic that went into deciding that an alarm occurred. There are basically three possibilities: a duplicate alarm logic database can be maintained at the workstation; the operator can somehow query the field panel to find out what logic was used; or the field panel can indicate the relevant alarm logic information at the time of the initial report. All of these variants can be found in commercially available systems.

Tabular reports can also be used to display information relating to energy metering. Such reports may include steam, hot water, chilled water, and electricity. Often the storage format of the data is such that the database may be directly accessed using one of the popular spreadsheet programs, which can then be used to perform all manner of statistical analyses and manipulations, including presenting the data graphically.

Many systems allow the operator to set up both short- and long-term logs of specific points or combinations of points. These reports can be used to keep track of trends over a period of time, such as the behavior of the space temperature in a particular building zone or a record of stop/start transitions for some piece of equipment. Most DDCs can be programmed to log equipment run time and notify an operator when a limit has been reached at which time the equipment should be considered for preventive maintenance.

- *Graphics.* On the basis of the well-established theory that a word is worth 1/1000 of a picture, various types of graphics have earned a prominent place in the typical workstation OMI. Two types of displays are particularly common. The first is the *x-y* plot or graphical trend log. The abcissa, or *x* coordinate, is usually time.

Fig. 5.1 shows an example of some electrical data taken from a database of archived meter readings. In this case, the *x* axis represents a full 24-hr day. Needless to say, color is essential in order to differentiate multiple plots on the same axes! In its absence, usable plots of single variables can be made or you could experiment with different line types made up of various combinations of dashes and dots. Plotting software may also be available to generate ad hoc plots of any accessible point from any field panel, singly or in combination with related points. Such software also usually allows you to set the sampling interval and length of the plotting period and to save the collected data in a disk file for later review. Such capabilities can be enormously valuable when trying to troubleshoot building system/controller problems that may be intermittent or where several interrelated control processes are involved.

The second prevalent type of graphic display is the building system schematic. Figure 5.2 is an example. Again, the main feature of such displays is the ability to view a large amount of related information without having to know the details of how to access the individual sensor, actuator, or software points that comprise the display. Figure 5.2 illustrates one of the great advantages of digital control. Not only can you see what the various sensors are reading, you can see what they are *supposed* to read,

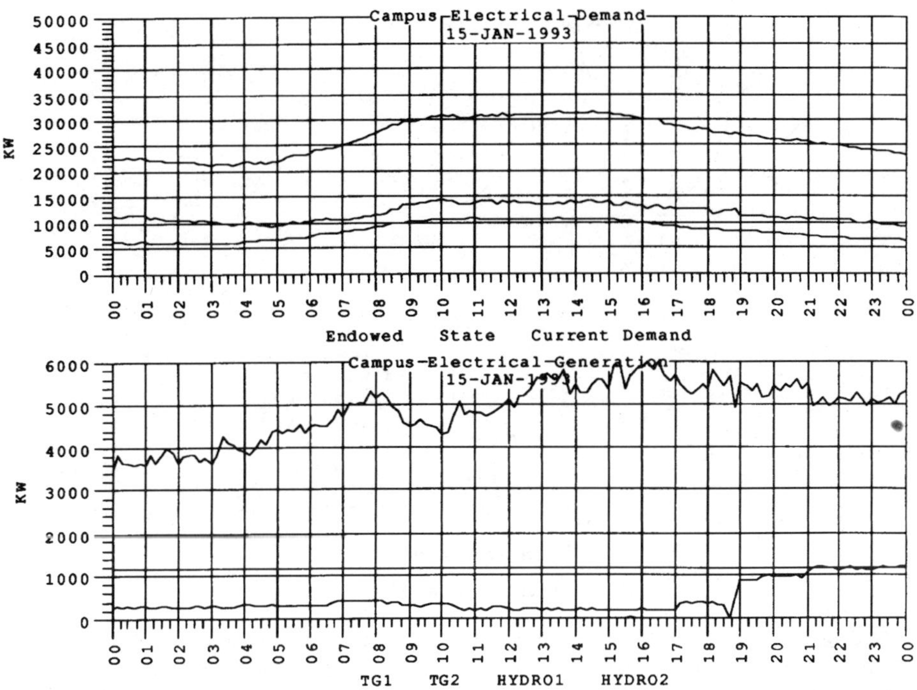

FIGURE 5.1. Twenty-four hour *x-y* plots of electrical demand and production. Such plots are easily produced from stored data with commercially available spreadsheet software packages. The need for color is obvious!

i.e., what the setpoints are! This is something that is just not possible with other control technologies.

An important adjunct to schematic displays is the graphic display composer program. The early ones were quite hideous. First you had to lay out your display on graph paper, then figure out the *x* and *y* coordinates of the end of each line, the handle point of each desired symbol, and the location of each variable. Then you had to enter the information by typing in all the coordinates using a text-oriented composer program, a process that was incredibly easy to foul up. Eventually you could "run" the graphic and see how close everything came to being where you wanted it to be. Today's programs make the process nearly child's play. Lines, symbols, and variables can be placed with the click of a mouse button and what you see is what you get.

Most graphic composer programs come with extensive libraries of predefined symbols representing the most common building system components. In Fig. 5.2, for

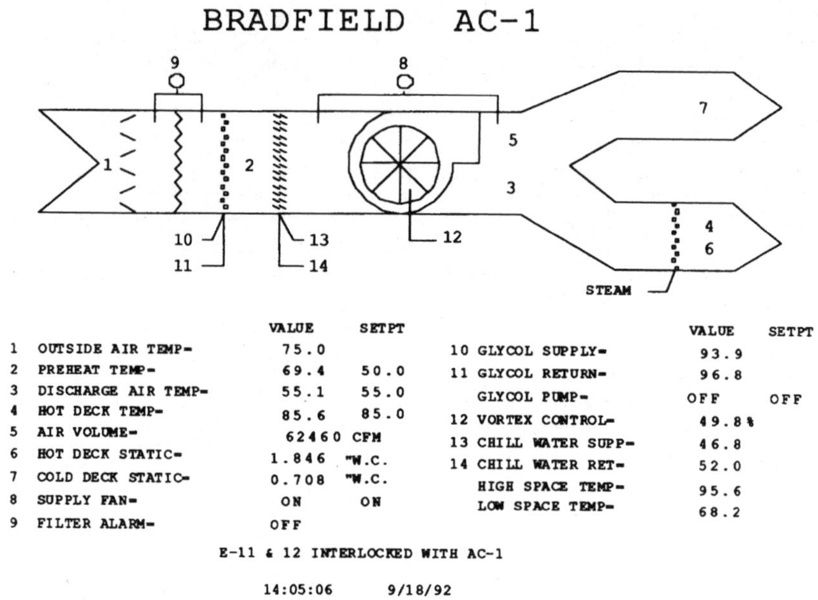

FIGURE 5.2. A typical real-time schematic diagram of an air handler in one of our buildings at Cornell University. Again, color would be an asset. One of the great advantages of DDC is that you can read both the current values of variables and their setpoints.

example, the dampers, filter, heating coils, cooling coil, and fan were all taken from the symbol library. You can also make your own symbols or add entire schematics to the library for reuse when a similar system is encountered later on.

Again, color can be used to good advantage. Variables that are normally shown in green, for example, can be configured to change color to red when they are above, or blue when they are below, certain threshold values. Alarm points can be made to flash in red when the associated alarm event is detected.

- *Command Interface.* Another obviously important OMI function is the ability to adjust or control binary or analog field points as desired by the operator. The adjustments may need to be short or long term. Several different approaches have been tried. The traditional approach is the typed-in command such as:

  SET FAN1 /STATUS=ON

  or

  ADJUST FAN1.DISTEMP.SETPOINT=58.0 /PRIORITY=OVERRIDE

  This is awkward, of course, because the operator has to remember the command, has to know how to reference the desired point, and has to know whatever other syntactical information is required to properly specify the desired action.

Another approach is to present the operator with menus. As an example, one menu might provide a listing of all controllable points, arranged perhaps by location, type, and function. A second menu might contain a list of actions appropriate to the type of point previously selected. If a value is needed, a dialog box could pop up giving the operator a chance to type in a value, or the current value could be displayed along with up and down arrows, which could be activated by the mouse cursor to cause the displayed value to increase or decrease. Once the operator is satisfied, a screen button could be clicked that would cause the OMI software to generate the appropriate command for transmission to the remote DDC device.

Yet another variation, common in the industrial process control world, is to make the schematic system displays interactive. To "manually" adjust a valve, for example, the operator positions the mouse cursor over the valve and clicks. This then causes an appropriate dialog box to appear allowing the operator to enter or select the desired valve position value.

- *Control Loop Tuning*. Although the trend is clearly in the direction of automated or adaptive control loop tuning, i.e., leaving it to the software in the field panel to figure out how to adjust the control parameters for smooth, stable loop operation, the OMI may have capabilities that permit an operator to determine the optimum tuning parameters remotely. One such program displays all the relevant loop gains and setpoints in real time on a single screen (reading the point access information from a disk file set up for this purpose) and allows the operator to perform an open-loop step test by giving him the chance to manually control the output device from the workstation. The computer then logs the effect of an operator-introduced step change in the position of the output device (valve, damper, or variable-speed drive) on the controlled variable (temperature, pressure, or whatever) and performs a computation that attempts to determine the optimum sampling interval and tuning parameters. I used the word *attempts* because occasionally the results of the test may be inconclusive if something occurs in the field that skews the results. If this happens, the test may have to be repeated a few times and the results averaged.

Some additional, perhaps more futuristic OMI applications, will be discussed at the end of this chapter.

5.5 MULTIVENDOR OMIs

One of the great attractions, at least for me, of a standard DDC communication protocol is the possibility of OMIs capable of simultaneously displaying data from the field panels of different manufacturers. Even today, however, there are some brave souls who have had some success not with common protocol, but rather with multiprotocol, workstations. I would like to briefly describe one approach to this problem because it illustrates the use of a particular way of referring to field data that is proving to be increasingly popular.

The problem, in essence, is that each DDC system refers to its hardware and software points in unique, proprietary ways. Moreover, the communication subprograms, or "drivers," are invoked using different numbers and kinds of parameters. What is needed is a method by which OMI applications can themselves refer to individual DDC points in a vendor-independent manner. The framework for one such a method is shown in Fig. 5.3.

From the point of view of applications, a common naming convention is required. One such convention is to construct a name from one or more components, each of which has relevance to the building system installation (as opposed to the naming requirements of the individual DDC systems). One possible scheme is to build the compound name from the building or facility in which the HVAC system is located, the name of the building system itself, and, finally, the name of the desired point. As

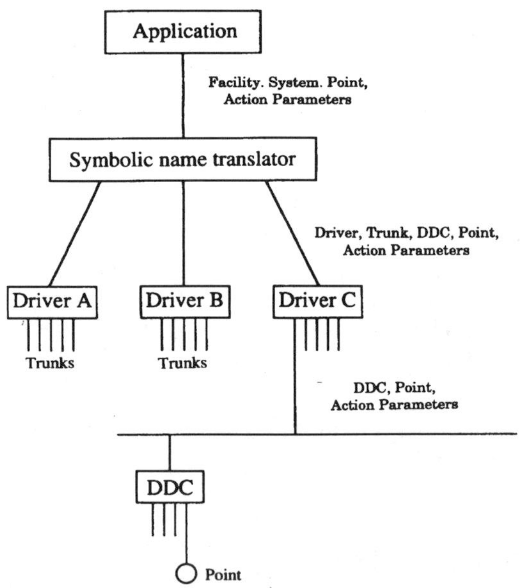

FIGURE 5.3. The Facility.System.Point naming convention, along with a name database and translation software, isolates building automation applications from the details of how a particular point is actually connected to the DDC network.

an example of this FACILITY.SYSTEM.POINT (FSP) naming method, consider a name like

CLARK_HALL.AHU1.MA_TEMP

The facility is Clark Hall, the system is Air Handler Unit 1, and the point is the mixed air temperature. The convention may be easily extended to accommodate areas within the facility, and subsystems within a system. For example,

CLARK_HALL.BASEMENT_MER.AHU1.CW.STEMP

would refer to the supply temperature of the chilled water subsystem of the air handler located in the basement mechanical equipment room.

The next aspect of the technique is the development of a database containing the parameters necessary to communicate with the field panel containing the point of interest. The parameters typically include the protocol or name of the driver to be invoked to carry out the communication, the particular hardware trunk or communication port to be used, the name or address of the DDC panel on that trunk, and whatever data are needed to identify the target point to the field panel in which it resides.

Once the database is set up, applications can simply call the symbolic name translator function with the symbolic FSP name and the action to be taken, such as READ or WRITE. It is then up to the translator to find the database entry corresponding to the supplied FSP and call the appropriate driver with the necessary parameters.

Although the FSP naming convention seems sensible even if there is only one driver to deal with, the greatest benefit of this approach is to be found in multivendor situations.

5.6 FUTURE OMI POSSIBILITIES

In one of the Star Trek movies, Scotty is appalled to find that the twentieth century computer he is trying to use doesn't respond to his polite request to start computing! Indeed the use of both voice recognition and speech synthesis are bound to become more prevalent in the years ahead. They are, of course, no panacea. There will still be times when selecting items from a menu makes more sense, particularly for an operator who is learning how to operate a new system and may not be sure of what all his options are or simply for those who prefer quiet!

Other input devices may also become more popular. I know people who hate mice, for example, and I am not referring to the fuzzy kind. Touch screens may be an answer for these individuals although I have always been averse to them because I can't stand fingerprints on my CRT. A compromise might be a touch sensitive pad where the fingers' motion over the pad drives the screen cursor and pressure anywhere on the pad acts like a mouse click. No matter what happens with I/O devices, the workstations themselves will blow the socks off today's machines in terms of raw power.

On the software side, we can expect far more integration of graphics and databases. It should be possible to marry fourth-generation query languages, databases of field data, and graphics to permit the answering of all sorts of ad hoc questions that a facilities manager might want answered. For example, let's see a plot of the daily energy use of the building that used the most energy last month. What was the average cooling energy use per square foot of all the buildings built before 1980 as compared with those built after 1980 during this past summer as contrasted with the summer before? And so on. The possibilities are endless.

Another area where advances can reasonably be expected is in the use of knowledge-based engineering to develop programs to diagnose building system problems. The diagnostic programs will probably take the form of *expert systems.* An expert system is a computer program that can provide solutions to problems that rival the solutions that a qualified expert, or experts, would be most likely to provide. Unlike programs that rely on fixed computational algorithms and formulas, expert systems depend on a "knowledge base" consisting of empirical facts and rules that can be used to draw inferences from the presence, in an actual situation, of a particular set of these facts. Expert systems have already been successfully developed for medical diagnosis, geologic exploration, and teaching, to name only a few areas.

The key to developing an expert system is to develop the knowledge base. Imagine, for example, the diagnostic power you would have at your disposal if you could "pick the brains" of the building architect, mechanical engineer, each contractor, subcontractor, and installer, the programmer who wrote the DDC operating system, the programmer who wrote the DDC application program, the maintenance tradesmen, all the EMCS operators, and so on. It is precisely in the area of how best to codify and store the accumulated knowledge, insights, and indefinable intuition of such experts that represents the current frontier.

Assuming that it *is* ultimately possible to produce a knowledge base containing all the relevant knowledge about specific building systems in specific buildings, the expert system should be capable of analyzing alarms and other "off-normal" events and providing the operator with (1) the most likely causes and the relative likelihood of each, (2) an explanation of the rules and facts used to arrive at each conclusion, and (3) recommendations for possible actions to be taken to resolve any uncertainties that may remain.

As with so many other computer capabilities, expert systems will never solve all the problems of the world, but they will certainly be fun to develop and work with!

BIBLIOGRAPHY

Bahill, A. T., *Verifying and Validating Personal Computer-Based Expert Systems*, Englewood Cliffs, NJ: Prentice-Hall, 1990.

Brothers, P., Knowledge Engineering for HVAC Expert Systems, *ASHRAE Transactions*, Vol. 94, 1988.

Brothers, P., and Cooney, K., A Knowledge-Based System for Comfort Diagnostics, *ASHRAE Journal*, September 1989.

Hayes-Roth, F., et al., *Building Expert Systems*, Reading, MA: Addison-Wesley, 1983.

Sell, P., *Expert Systems—A Practical Introduction*, New York: Wiley, 1985.

Chapter 6

Communication Systems

SUMMARY

Stand-alone computers, by definition, are not concerned with data communications. But the moment we need to connect two or more unitary controllers, DDC panels, central computers, or operator workstations, intersystem communication becomes crucial.

Data communications is the process of conveying information, expressed in some predefined binary format, from one piece of computer-based equipment to another. The set of codes, message structures, and procedures, both in terms of hardware and software, which permits this communication is referred to as the communication *protocol.* The possible variations and permutations of the elements which comprise a protocol are almost limitless. For this reason, standards are essential if systems from more than one vendor (or from the same vendor!) are to be able to exchange information.

One of the most useful developments in the worldwide effort to achieve communication standards has been the International Organization for Standardization's (ISO) "Open Systems Interconnection (OSI) Basic Reference Model." The model represents an effort to make the communication problem manageable by breaking it down into a series of smaller problems, each of which can then be tackled independently of the others. This has been accomplished by defining a hierarchy of functions arranged one on top of the other in seven "layers." See Fig. 6.1.

In trying to understand data communications, it is essential to distinguish between the "languages" used to program computers, discussed in Chapter 4, and the communication "languages" used for information exchange. They are totally separate entities. A computer must understand both to carry out its monitoring and control functions and to communicate with other parts of the system.

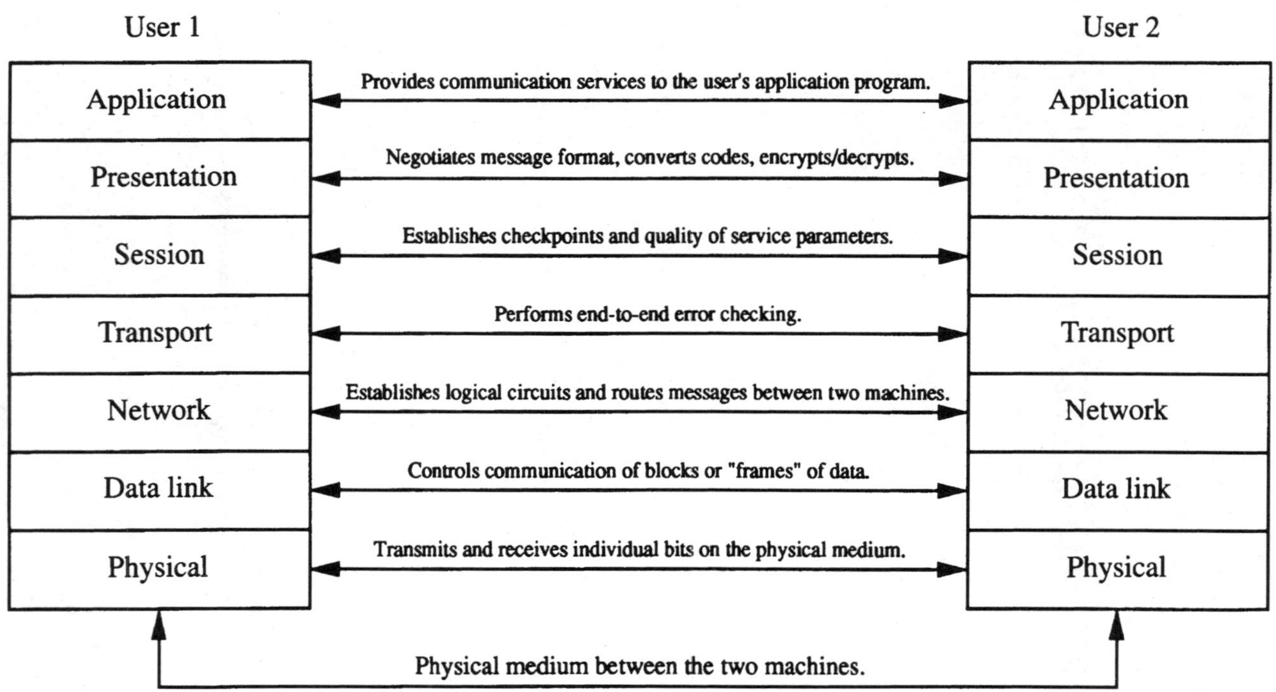

FIGURE 6.1. The ISO's Open Systems Interconnection Basic Reference Model establishes a layered hierarchy of functions that forms the basis for the development of most modern data communication standards. Each layer is implemented using a different protocol, and the sum of all the protocols in a particular implementation is called the *protocol stack*.

6.1 COMMUNICATION REQUIREMENTS

As I have pointed out, not all DDC systems need to communicate: stand-alone controllers play a reasonable and important role in many cases—that of replacing analog electronic or pneumatic controllers of similar function more or less on a one to one basis.

So when do communication capabilities become important? There are two fundamental situations. First, when it is necessary to exchange information to carry out some distributed control strategy. A very common example is the sharing of outside air temperature among several controllers so that each one does not have to have its own sensor. Another example would be sharing the on/off status of fans or pumps to effect an interlock function in software.

The second fundamental situation is when we need to centralize or consolidate some computing function. The most common example is the operator workstation. To be able to display alarms, graphics, utility metering information, run time reports, or perform any of the other OMI functions discussed in the last chapter, the workstation needs to be able to "talk" to the various DDC computers to retrieve or read the necessary data.

Okay, we may need to network our controllers. But what are some of the things that need to happen for two or more computers to be able to talk? Let's look at some of the factors that need to be considered in building a computer network, and then we will look at how the ISO's communication model helps to bring order to the chaos.

First, the computers need to be somehow physically connected. The connection may be temporary, such as via a dial-up telephone line, or permanent using twisted pair wires, coaxial cable, or fiber optic cable. Even air could be the physical medium if we were communicating using radio or infrared. Then, what is the nature of the signaling over our chosen medium, i.e., what constitutes a binary zero and what constitutes a one? How are the computers to get synchronized so that they all know where one message starts and another one ends?

Next, how is access to our connection medium arbitrated? Obviously not all the computers can talk at the same time! We need to agree on a common *network access method*, sometimes also called *medium access control* method.

Once orderly access to the medium has been established, we need to be able to "address" a message from one computer to another. How do we know which DDC device is which?

We also need to make sure that all messages are received error-free. This means that we need some reliable mechanisms to (1) detect errors and (2) recover from errors once they are detected.

How can we keep one computer from overwhelming another with requests or data, i.e., how can we control the data flow?

Assuming that all the preceding questions are resolved, what is the structure of the actual messages themselves? The situation is very much like that of the international telephone system: it is relatively easy to call anywhere in the world; but *communication* only occurs if both parties to the call speak the same language!

Given all the combinations and permutations of the possible answers to the above questions, it is not surprising that getting equipment from different vendors to communicate has been a bit of a challenge. Each vendor of DDCs, or computer equipment in general, has had his own, usually proprietary, solution. Fortunately, the need for data communication standards has been widely recognized, even in the building controls industry! Not everything is cut and dried, however. A friend of mine used to say, "the nice thing about standards is that there are so many of them!" While this cynical view is certainly true, some standards are destined to play a greater role than others. One such standard is the ISO's Open Systems Interconnection Basic Reference Model. It is important because it gives us a common way to group the various related problems together so that we can work on one set of issues at a time, knowing that when we are done the pieces will fit together.

While the primary purpose of OSI is to give data communications engineers a framework within which to develop standards, the model also provides an excellent way to coordinate the study of data communications technology. We will use the layer concept to introduce the major ideas needed to understand data communications and then we will look at DDC communications, both in terms of the state-of-the-art and what the next several years are likely to bring.

6.1.1 The OSI Layered Architecture: A Mail Analogy

Before diving into the details of each OSI layer, let's look at an analogy between computer messaging and sending mail (the traditonal, nonelectronic way). The analogy will show that conceiving of, writing, packaging, addressing, and sending mail is also a process with a multiplicity of discrete component parts. The analogy makes the point that the extent of the name and address information that needs to be provided depends on the nature of the recipient and how complex the route is from sender to receiver.

6.1.1.1 Application Layer We begin with a need to communicate. The need may involve a question, an answer, or the desire to make a statement or send a greeting. You may want to communicate with anyone who will listen or with only a specific recipient. Let's say you want to write a letter to a business associated named Fred and tell him about this great DDC book you have been reading. As in the computer case, the amount of name and address information you need to provide depends, roughly speaking, on how big a group Fred belongs to and how far away he is. "Fred" may suffice for someone in a small office, but a last name may be needed if Fred works in a big company. If Fred is self-employed and there is no one else at Fred's address, you might even be able to skip the name on the envelope entirely and just send the mail to Fred's office address.

A program accessing the communication services of the application layer has a need to communicate that is specific to the function of the program, i.e., application specific. Let's say we have a DDC program that wants to notify interested parties that a temperature alarm has occurred. The notification could be broadcast throughout the network or only sent to a particular recipient. Though it may suffice to send the notification to a particular computer (Fred's office), it may need to be sent to a particular process within the computer (Fred), such as the one that rings bells and flashes lights, and thus has a particular name (application identifier).

6.1.1.2 Presentation Layer Having formulated the idea to be contained in your letter, the next thing you will need to do is decide what language to use. This may, of course, be assumed. But it could also be that you and Fred both know more than one language or you may have a secret code, or at least certain code words, that you like to use.

The alarm message needs to be encoded in a particular way so that the receiving application will understand it. The encoding method may be agreed to implicitly in advance or there may have to be a negotiation concerning which format to use. In addition, bandwidth or security requirements may dictate that the message be compressed or encrypted.

6.1.1.3 Session Layer You will need to decide on the "quality of service" you require to convey the message. This may lead to the preferential selection of Federal Express or United Parcel Service over the Postal Service (or vice versa). In the latter case, you may opt for airmail, first class, book rate, or special delivery.

The alarm message may be sent with any one of several possible priorities which determines how quickly the message is expedited through the network. In addition, certain message paths may need to be avoided because of high traffic loads or exceptionally noisy circuits.

6.1.1.4 Transport Layer You may decide that you would like to know that Fred actually got your mail. You could, of course, wait for Fred's response but you might also wish to use certified or registered mail as further assurance. (Keep in mind that certified mail only has to be signed for by *someone* at Fred's address while registered mail has to be signed for by *Fred* himself). Note, however, that only by getting a response directly from Fred that deals with the substance of your letter can you be sure that he actually read and understood your message.

The transport layer sends end-to-end acknowledgments assuring that the message at least got to the correct machine. As with Fred, a transport layer acknowledgment does not mean that the recipient application has received and correctly interpreted the alarm message, let alone taken any action based upon it.

6.1.1.5 Network Layer You may need to supply information on the country, state, and city where Fred is located, but the route taken by your mail is almost always out of your control, and mail sent by different carriers will almost certainly travel by different paths. If you leave out any of these items, the letter carrier will generally make assumptions. For example, if you leave off the state and country, the carrier will assume the city you have specified is in your own state and country.

A network layer address is only needed when the two computers are not on the same local network. In this case, the network layer protocol determines the best route for the message to take. This may involve several hops, but the local network layer usually only needs to know how to direct messages to the nearest intermediate destination (the next post office or sorting point). The network layer address also provides for the possibility that two computers have the same local address (two Fred's on Main Street in Norwalk, but one in Connecticut, one in Ohio).

6.1.1.6 Data Link Layer You next have to determine how your letter is going to enter the mail system. You may have to go to the post office or drop the mail in a designated slot or box. If you put it in you own mailbox, you may need to know to raise the outgoing mail flag, otherwise the carrier may ignore it. You may also have to supply a local address containing, for example, Fred's street and apartment number.

The data link layer provides local message conveyance. Each data link has specific rules by which messages may be sent, i.e., how the network is accessed. The data link address applies only to a particular computer on the local network (it assures the mail gets to Fred's office, but not to Fred).

6.1.1.7 Physical Layer Finally, you need to decide how the symbols that make up the words that form the ideas in your letter are physically constructed. Will you use pen, pencil, typewriter, or printer? Also, there is the question of how the mail is to be physically conveyed. Will your selected service use truck, bus, car, ship, or plane?

The physical layer determines how the zeros and ones that form the computer message are actually signaled. It also dictates by what medium they are conveyed, e.g., twisted wire pairs, coaxial cable, or optical fiber.

Other analogies are also possible. If you think about it, telephone communication likewise bears many similarities to computer communication. In any event, the OSI layer concept should not intimidate you—it is simply a model—and its utlity in organizing data communication concepts should become clear as we proceed.

6.2 THE PHYSICAL LAYER

According to OSI, the purpose of the physical layer is to provide "mechanical, electrical, functional and procedural means to activate, maintain and de-activate physical connections for bit transmission" between *open systems*, the OSI term for computers that adhere to OSI standards. Physical layer concerns are therefore what kind of cables and connectors to use; what sort of electrical signals should be put on the cables to represent zeros and ones; and how the physical connections are to be established. To put it in English, the physical layer deals with getting bits from the infamous point A to the likewise notorious point B!

Before describing the many ways used in DDC systems to accomplish this feat, let us recall that there are only two ways to convey a string of bits: in parallel, i.e., many simultaneously, or serially, one after the other.

To transmit bits in parallel, DC voltages corresponding to the value of each binary digit are applied simultaneously to the conductors of a multiconductor cable. A pulsed control signal called a *strobe* is applied to another conductor to tell the receiver that data are present on the signal lines and ready for capture. If the data port uses transistor-transistor-logic (TTL) levels, for example, a 1 bit can be represented by 5 VDC, while a 0 bit corresponds to 0 VDC. When the data have been successfully gated into the receiver, the voltage on yet another conductor is raised to acknowledge receipt. The parallel printer interface on personal computers works this way.

Most commonly, parallel data transmission is confined to internal computer data buses, as discussed in Chapter 3. In this case, additional conductors are required to address the location to which data are to be transmitted although the data lines (or data bus) may double as address lines with some reduction in throughput. The primary advantage of parallel transmission is speed: hundreds of megabits per second data transfer rates are readily attainable. The main disadvantages are the cost and practical awkwardness of multiconductor cable.

In the real world of HVAC DDC, where a time scale of seconds, or even minutes, may be acceptable for most monitoring and control purposes, data transmission between devices is serial. In this case, each bit appears at the output of the computer or terminal device as a specific DC voltage or voltage transition. In data communications, computers and terminals are referred to as *data terminal equipment*, or DTEs.

If the voltage signal is coupled directly to the physical medium, we speak of *baseband, DC*, or *digital* transmission. Because the electrical characteristics of cables tend to attenuate and distort the waveforms, it is sometimes desirable to amplify the signal before applying it to the cable in order to extend the communication range. A device that amplifies the DTE signal, or converts it to a current loop signal, is called a *line driver* and almost always requires DC continuity from end to end.

If the data signal is used to alter, or "modulate," an analog, continuously varying sinusoidal signal that serves as a carrier of the digital information, we speak of *broadband, AC*, or *analog* signaling. The device that performs the *mo*dulation or *dem*odulation is referred to as a *modem*. Because AC signals can pass through transformers and can be easily amplified, end-to-end DC continuity is *not* required, a fact that enables some types of modems to be used on ordinary, public telephone circuits. Line drivers and modems are referred to generically as *data communications equipment,* or DCEs.

6.2.1 Baseband Signaling

Two types of baseband signaling are worth mentioning because of their widespread use. The first is called *nonreturn to zero* or NRZ. NRZ simply means that the DC signal level applied to the medium retains its value for the entire length of the associated binary zero or one. This works well for short streams of bits such as are generated by the serial asynchronous communication devices built into most video display terminals and low communication speed DDC equipment. In asynchronous transmission there is no fixed time relationship between one byte and the next.

The basis for asynchronous communication is often an IC called a *universal asynchronous receiver transmitter* (UART). UARTs contain shift registers into which data bits are strobed, in parallel, one byte at a time. The UART then shifts out the data sequentially, one bit at a time, by applying the appropriate voltage to the communication line. The rate is determined by the sender's clocking circuits. In order for the receiver to know where a byte begins, the UART first puts out a start bit, usually a binary zero. This works because the line is held at the voltage corresponding to a binary 1 between transmissions. At the end of the byte, the UART tacks on a stop bit, usually a 1.

NRZ depends on the assumption that the transmitter and receiver have clocking circuitry which is running at the same speed. If the transmitter and receiver don't agree on how long a bit is, such as might happen if someone inadvertently set the UARTs to different baud rates, the receiver won't "see" the same number and value of bits that the transmitter sent, resulting in chaos. In the case of synchronous, bit-oriented protocols, where long streams of bits are sent with no intervening start and stop bits, NRZ encoding can run into trouble if the data contains a run of all zeros or ones. In this case, even small differences in clock timing could cause the receiver to lose track of its position in the bit stream and eventually gain or lose a bit.

To overcome this problem, self-clocking codes were developed. One the most popular is called *Manchester encoding* which is used by the Ethernet LAN that we will discuss later. In Manchester encoding, a zero is signaled by a one-to-zero transition in the middle of a bit period. A one is indicated by a zero-to-one transition. The presence of at least one level transition per bit time essentially eliminates the clocking errors that can occur with NRZ. See Fig. 6.2.

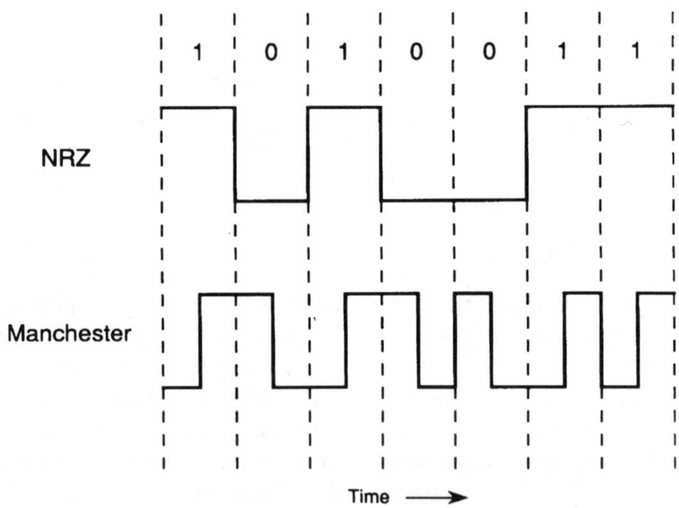

FIGURE 6.2. Nonreturn to Zero (NRZ) and Manchester encoding of binary digits. The latter is referred to as a self-clocking code because there is a voltage level transition during every bit time.

6.2.2 Modulation Techniques

Modems are used so frequently in DDC installations, particularly to provide remote dial-up capability, that it is well worth understanding how they, and modulation techniques in general, work. Binary information can be transmitted by modulating any of the three characteristics of a sinusoidally varying AC signal: its amplitude, its frequency, or the phase relationship between successive cycles. See Fig 6.3. A modem may employ any one or a combination of these techniques. For low-speed modems (300–1200 bits per second), the most common technique is to vary the frequency of an audio tone. This is called *frequency shift keying,* or FSK. The resulting signal can be transmitted on commercial voice-grade telephone lines or inexpensive pairs of wires within a frequency bandwidth of only 3000 Hz.

As examples of commercially available modems, let's look at the Bell 202 leased line modem and the Bell 212 modem for the public switched (dial-up) telephone network (PSTN). (The Bell System has an open-ended classification scheme for their

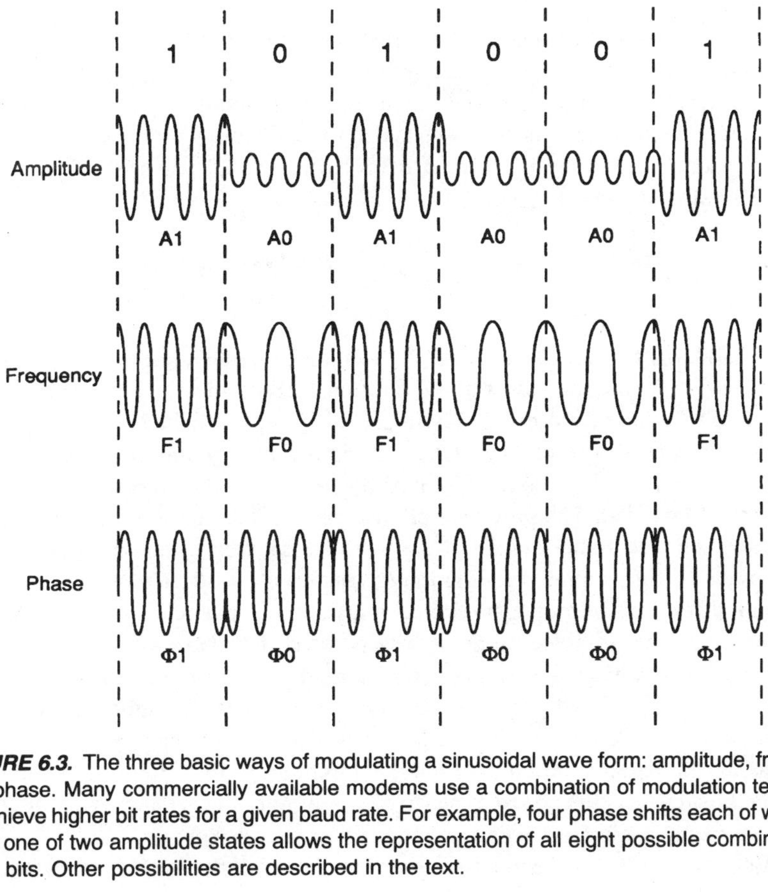

FIGURE 6.3. The three basic ways of modulating a sinusoidal wave form: amplitude, frequency, and phase. Many commercially available modems use a combination of modulation techniques to achieve higher bit rates for a given baud rate. For example, four phase shifts each of which can be in one of two amplitude states allows the representation of all eight possible combinations of three bits. Other possibilities are described in the text.

modem specifications with all modems in a given 100 series sharing common attributes. Bell 200 series modems, for example, are for use with voice bandwidth communication channels, speeds up to 9600 bits per second (bps), serial data transmission, and digital input signals.)

The Bell 202 modem uses FSK with a frequency of 2100 Hz representing a binary zero and 1300 Hz representing a binary one. The maximum bit rate is 1200 bps on unconditioned lines. Since only one pair of frequencies is used, the communication can only be full duplex (two DTEs able to talk simultaneously) if separate wire pairs are used for transmitting and receiving. If only a single pair is available, the DTEs are forced to alternate, a mode of communication referred to as half-duplex.

Bell 202 modems can be operated in both point-to-point or multipoint modes. Point-to-point means that only two devices are interconnected. In a multipoint configuration, several DDC devices can hang off a common trunk. To control which machine can transmit at any particular time, the master/slave network access method is usually used. See Section 6.3.1. Because the frequencies used are VF, i.e., in the voice frequency band of 300 to 3000 Hz, these modems can be used on ordinary leased phone lines. They can therefore can be used to link facilities over a wide geographic area.

Note that several types of leased lines are available from the telephone companies. A *local area data circuit* (LADC) is simply a pair of copper wires and must be used where line drivers or other devices are used which require DC continuity. The distance over which LADCs can be used is determined simply by the power loss they introduce, usually measured in dBm. A dBm is the decibel power level ratio referenced to 1 milliwatt (mW):

$$\text{dBm} = 10 \times \log_{10}\frac{P(\text{mW})}{1(\text{mW})} \tag{6.1}$$

Most VF modems transmit at 0 dBm. If the threshold sensitivity of the receiving modem is −16 dBm and the power drop is around 4 dBm/mile, the distance attainable with the LADC would be 4 miles. These are fairly typical values.

Another common type of leased line is simply an unconditioned voice-grade phone line. Such lines may be readily used for low-speed data transmission over as wide an area as needed. The guiding Bell system specification for such lines is Bell 3002. It specifies electrical characteristics such as frequency bandwidth, maximum distortion due to signal strength variation (attenuation distortion), maximum distortion due to differences in propagation time at different frequencies (delay distortion), and minimum signal-to-noise ratios. A key point is that these circuits are not DC continuous; they use amplifiers to maintain a maximum signal power loss of 16 dBm from end to end. This is why they can be used over essentially unlimited distances.

"Conditioned" voice-grade phone lines can also be obtained. Such lines have tighter specs, particularly in terms of the maximum allowable attenuation and delay distortion. Fortunately, the added expense of conditioned lines is rarely required due to improvements in modem design.

The Bell 212 modem is designed for point-to-point full duplex communication on the PSTN, also at the relatively lethargic speed of only 1200 bps. As part of the handshaking sequence that occurs when one of the two cooperating modems answers the phone, one modem assumes the role of originator and transmits on a carrier frequency of 1200 Hz. The other modem then goes into answer mode at 2400 Hz. To encode zeros and ones, the Bell 212 specification uses four distinct sine wave phase changes to indicate the four possible binary transitions: 0 to 0, 0 to 1, 1 to 0, and 1 to 1. This is called *four-level differential phase shift keying*. Since a phase change occurs once each bit time, the receivers are able to easily maintain synchronization. The nice thing about differential PSK is that the receiver does not need to have a knowledge of the absolute phase of the incoming signal (relative to some reference signal). Only large phase shifts need to be sensed to recreate the original binary data stream.

In general, higher transmission speeds demand greater bandwidth from the transmission medium. But are we squeezing the most information we can into our available bandwidth with simple FSK or PSK? Information theory shows that the number of state transitions per second, e.g., from low frequency to high frequency and vice versa, is directly proportional to the bandwidth required to convey these state transitions. The number of such transitions per second is called the *baud rate*. If each transition represents a bit changing state, the baud rate is identical to the transmission rate expressed in bits per second. To transmit more data within a given bandwidth, we must therefore come up with ways of making each transition *represent the state of more than one bit* thus increasing the transmission rate in bits per second while keeping the baud rate, and hence the required bandwidth, the same.

This is done with multimode modulation techniques. For example, instead of sending an audio signal with just two frequencies, four frequencies might be sent, each frequency correlating to the state of not one but two bits in the modulating data stream. In this case, the number of bits per second transmitted is equal to twice the baud rate. One commercially available technique uses 12 different phase transitions, 4 of which are also amplitude modulated. This gives 16 different states, the proper number of states to represent all possible combinations of four data bits. At 2400 baud, such a modem is actually conveying information at the rate of 9600 bps. The trade-off is that such modems require more sophisticated circuitry and cost more than simple low-speed, bi-state modems. Happily, the prices are declining such that 9600-baud modems are now about the same price as 1200-baud modems were in the 1980s!

To ensure reliable data transfer, several data link protocols have been established specifically for use with high-speed modems such as the Microcom Networking Protocols and the CCITT's V.42. These protocols provide for error detection and retransmission at the modem level without having to involve their respective DTEs.

6.2.3 Physical Media

It used to be that selecting a particular DDC vendor was the same as picking a networking technology; this, in turn, tended to fix the choice of physical medium. Today, more and more vendors are making use of standard local area network (LAN) technologies that are capable of running over multiple media. The most common media are twisted pairs, coaxial cable, fiber optic cable, and existing power lines.

From an application point of view, the main questions are: how fast, how far, and at what cost? From an engineering perspective this comes down to: which cable will yield acceptable attenuation and signal distortion at the frequencies of interest over the distance required and where will the cable be used, i.e., what kinds of mechanical, thermal, and radiational stresses will the cable have to withstand?

The suitability of a cable for a particular environment, independent of its electrical characteristics, is determined by its insulation and jacketing material. Depending on the fire code in use, some materials, such as certain polyvinyl chlorides and Teflon, are rated for use outside of conduit and may be run over drop ceilings or in plenums, resulting in much lower installation costs. The catalogs of the major wire manufacturers, such as Alpha and Belden, are treasure troves of data concerning the pros and cons of different cable configurations.

6.2.3.1 Common Cable Characteristics The purpose of any physical medium is to convey the input signal as faithfully as possible to the output. Three things can happen to mess things up. First, the signal can get weaker. This is called attenuation. In metallic cables, this is mostly the result of ohmic or I^2R losses (where I is the current and R is the DC resistance). In fiber, imperfections in the core material can cause optical scattering with attendant power loss.

The second problem is signal distortion. In metallic cables, the main culprit is cable capacitance (C). As the voltage across the input rises, some of the energy first goes into establishing an electric field with neighboring conductors resulting in a sloped output signal. As the bit rate increases, the duration of a bit may approach the RC time constant of the cable. In this case, the output waveform "smears" to the point where the signal is no longer recoverable. The effect of capacitance is also frequency dependent, affecting higher frequencies more than lower ones (due to losses in the dielectric material between conductors), a factor of more importance in analog signaling.

Another cause of distortion are the reflections of voltage caused by impedance mismatches. Depending on how these interact with the impressed signal, bits can appear to be shifted in time or lost altogether.

In the case of fiber, the main cause of distortion is the dispersion of a bit caused by light rays that follow paths of differing length from source to receiver. See Section 6.2.3.4.

The third source of difficulty is noise, fluctuations in voltage caused by external electric and magnetic fields. Since cables differ in their immunity to noise, this can be a major selection factor. One of the greatest advantages of fiber, of course, is that it is entirely immune to electromagnetic disturbance.

The electrical characteristics of metallic cables are determined entirely by their geometry and their constituent metallic and insulating components. Impedance is the overall resistance of a cable to the flow of alternating current. An important attribute of a cable is its characteristic impedance, R_0. The theory of transmission lines shows that R_0 is a pure resistance which, if placed at the termination of such a line, will entirely absorb the arriving energy without reflecting any of it back toward the source. The choice of characteristic impedance is usually dictated by the input or output impedance of the equipment that is to be interconnected.

The capacitance of a cable varies with geometry and the dielectric constant of the insulation between adjacent conductors as

$$C \propto \frac{\varepsilon}{\ln (D/d)} \tag{6.2}$$

where $\propto$ means "proportional to," ε is the dielectric constant, d is the conductor diameter, and D is a characteristic distance such as the distance between conductors in a twisted pair and ln is the logarithm function to the base $\varepsilon = 2.718$. Equation 6.2 shows that the capacitance of cable can be reduced by choosing an insulation with a lower dielectric constant, by decreasing d, or increasing D.

Why not just select a cable with the lowest possible capacitance? The problem is the constraint that exists between R_0 and C:

$$R_0 = \frac{1}{VC} \tag{6.3}$$

where V is the velocity of electromagnetic propagation in the cable and C is the capacitance per unit length. V, it turns out, is strictly determined by the dielectric constant and the magnetic permeability of the insulation, the latter of which is close to its value in free space for practical insulators. Thus, once R_0 and ε are chosen, C is fixed by Eq. (6.3).

6.2.3.2 Twisted Pairs Pairs of copper wire, 20–26 AWG (American Wire Gauge), are still the most common physical communication medium for DDC. These pairs of wire are usually twisted together within their insulating jacket to nullify the noise-generating effects of varying magnetic fields and are thus referred to as *twisted pairs*. Because of the twists, the currents generated in adjacent pieces of the same conductor flow in opposite directions and tend to cancel out. See Fig. 6.4. Electric fields, on the other hand, affect both conductors equally and produce a *common mode* voltage, relative to ground, which must be taken into account by receiver circuitry.

Because electromagnetic fields are so commonly generated by the types of mechanical equipment found in the DDC environment, it is often desirable to shield the twisted pairs in a sheath of braided copper, aluminum foil, or both. Often an additional conductor, called a *drain wire,* is provided. The drain wire runs along the shield and is used to electrically connect the shield to its termination. Such shielding, properly grounded at a single point, can greatly reduce the electromagnetic noise that might otherwise impair communications. In general, the use of shielded cable is a relatively inexpensive form of insurance and is highly recommended, even if the cable is run in conduit!

The actual signal conductors may be either solid or multistranded wires. Solid conductors are needed if the cables are to be interconnected on telephone-company-

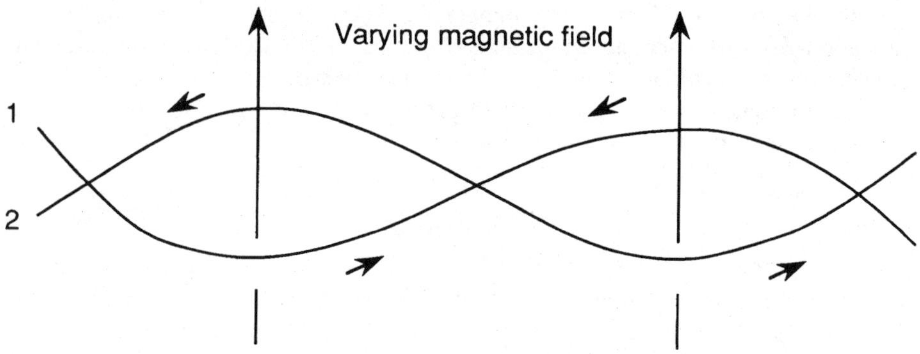

FIGURE 6.4. In twisted wire pairs, varying magnetic fields, indicated by the vertical arrows, induce currents in wires 1 and 2, indicated by the short arrows, that cancel out in adjacent twists thus effectively negating one of the major sources of electrical noise.

type M-66 or "punch-down" blocks. Stranded wire, on the other hand, is more forgiving in the sense that it is harder to nick, and later break, all of the conductors at a single junction (a real danger if you are more of a software type than a hardware type)! In spite of such perils, twisted pairs are by far the easiest medium to work with, usually requiring little more than a wire stripper and a screwdriver or punch-down tool.

For purposes of comparison later on, the characteristic impedance of a typical shielded, twisted pair is around 150 Ω; the attenuation in decibels per km (dB/km) ranges from 1.5 at 1 MHz to 6.0 at 10 MHz; and the capacitance/foot is around 30 pF.

6.2.3.3 Coaxial Cable Another common physical medium is coaxial cable. "Coax" consists of a single, central conductor surrounded by an insulating dielectric material, usually solid polyethylene or foamed polyethylene. The entire cable is then shielded as described above. Whereas twisted pairs have a practical frequency bandwidth of about 10 MHz, the bandwidth of coax can reach 1000 MHz. This makes coax ideal for conducting signals useful in the radio frequency (RF) spectrum, including both baseband and broadband LAN transmissions. "RF" is an old generic term that was originally applied to any electromagnetic signal capable of propagating in free space, the only known use for which was radio! Now the term is applied to signals starting in the tens of kilohertz and reaching into the gigahertz (10^9 Hz) region. Besides radio these signals encompass television, microwave, and radar signals but stop short of the infrared spectrum which begins around 10^{12} Hz.

Because of its relatively low loss at radio frequencies, coax has been, until the recent introduction of optical fiber, the medium of choice for military RF applications. Their specification, MIL-C-17, is still the basis for most commercially available coax and gives us the common "RG" prefix which stands for "radio guide" and the suffix "/U" which means the specification has "universal" applicability.

Three characteristic impedances are commonly available: 50, 75, and 93 Ω. Cables with these impedances represent trade-offs between power-carrying capacity, which peaks around 30 Ω, and attenuation which is a minimum around 77 Ω.

In practice, the LAN technology and its associated specifications usually dictate the choice of cable. The original "thick wire" Ethernet specification resulted in the mass production of coax 0.405 in. in diameter that differed from any previous milspec type. In contrast, the "thin wire" Ethernet specification allows the use of standard RG-58/U which is about 0.19 in. across. Both cables have a nominal impedance of 50 Ω. Another common LAN that specifies coax is ARCNET. It uses 93 Ω RG-62/U.

Although coax uses special connectors, it is nonetheless easy to install and fairly easy to work with, especially with a little practice.

6.2.3.4 Fiber Optic Cable Where coaxial cable acts as a waveguide for the relatively low-loss propagation of RF energy, optical fiber serves the same function for energy in the optical spectrum, which includes both visible and invisible light. In terms of wavelength, visible light extends from about 400 nanometers (nm) for violet light to about 700 nm for red light. A nanometer is one-billionth or 10^{-9} m. Since

$$f = \frac{c}{\lambda} \tag{6.4}$$

where f is frequency in Hz, c is speed of light in m/s ($\approx 3 \times 10^8$ m/s), and λ is wavelength in meters, the wavelength of violet light, for example, corresponds to a frequency of 7.5×10^{14} Hz, or 75,000 GHz. Light with a wavelength longer than 700 nm is called infrared. The vast majority of fiberoptic communications takes place at infrared wavelengths and is therefore invisible to the naked eye (in spite of all the nifty Madison Avenue ads showing bundles of fibers with light sparkling out the ends)! The most common wavelengths used in commercial equipment are 850, 1300, and 1550 nm.

Before presenting some of the most important fiber optic communication concepts, a few basic terms and ideas from optics need to be reviewed. The first is the notion of the *index of refraction* of a material. This is the ratio of the speed of light in a vacuum to the speed of light in the material. Since light attains its maximum speed in free space, this ratio, designated n, is always ≥ 1.0. For typical glass fibers, n ranges from 1.4 to 1.7.

The second idea is the law of refraction discovered by, and named after, Willebrord Snell in 1621. Snell's law states that a beam of light traveling through a medium with index of refraction n_1 and striking the interface to a second medium with an index of refraction n_2 at an angle θ_1 with respect to the normal to the interface will be bent or refracted such that it leaves the interface in the second material at an angle θ_2.

$$\frac{\sin \theta_1}{\sin \theta_2} = \frac{n_2}{n_1} \tag{6.5}$$

The situation when $\theta_2 = 90°$ defines the "critical" θ_1 also known as the Brewster angle.

Any light ray hitting the boundary between the two materials at an angle greater than the critical angle will be totally reflected back into the first material. This is the underlying principle of fiber optic transmission.

Fiber optic cables are made up of a core material, usually silica glass, and a surrounding cladding material which always has a lower index of refraction than the core. Light rays which enter the core at a sufficiently narrow angle with respect to the axis of the fiber will strike the cladding at angles greater than the critical angle and thus will always be reflected. This process is referred to as *propagation by total internal reflection.* The maximum angle with respect to the fiber axis that a light ray can enter the fiber and still be propagated defines a cone of acceptance and the sine of the maximum angle is called the *numerical aperture* (NA) of the fiber.

In common with metallic waveguides operating at radio frequencies, optical fibers permit only certain paths through the core to be traversed without massive attenuation. (This has to do with the fact that light rays that fail to travel an integral number of wavelengths between reflections tend to lose energy to the cladding due to the direction of their electric and magnetic field vectors at the point of reflection—a complicated subject!) These allowable paths are called *modes.* In general, the fatter the core and the larger the NA, the more modes that can propagate through the fiber. Fibers thus fall into two broad categories: multimode and single mode. Standard multimode core/cladding diameters (in micrometers) are 50/125, 62.5/125 and 100/140. A standard single-mode fiber is 9/125.

What are the practical differences between the two types of fiber? Multimode fibers are much easier to make and therefore are much less expensive. But the fact that light rays entering the fiber at the same instant can travel by routes of different length to the exit of the fiber means that light pulses get spread out in time, thus placing a very definite limit on the minimum pulse width that can be carried. This pulse width, in turn, determines the maximum frequency that can be propagated, i.e., the bandwidth or bit rate of the fiber. Since there is only one allowable path through a single-mode fiber, the bit rate is theoretically infinite.

There are, of course, distance limitations in real fibers. These are the result of several factors including impurities in the core which scatter some of the light into the cladding or simply absorb it; deviations from circularity of the core; and variations in the refractive index along the length of the fiber. There are also losses at each splice or coupling.

Besides the fiber itself, fiber optic links require light sources and detectors. The most common sources are laser diodes and light-emitting diodes (LEDs). Compared to laser diodes, LEDs are slower, produce less optical power for a given input current, have a relatively large emission angle and a broader spectral width. As a result LEDs tend to be used in lower-speed applications (up to 100 MHz) with large core multimode fibers (> 50 μm) over short distances (< 10 km). The most common detectors are also semiconductor devices, either positive-intrinsic-negative (PIN) diodes or avalanche photodetectors (APDs). The latter not only detect light but amplify the optical signal by a factor of 10 to 250.

At the present time, both analog and binary information is conveyed over fiber by intensity modulating the source of optical power, a process known as incoherent modulation because neither the frequency nor phase of the optical carrier is affected. The maximum modulating signal frequency is perhaps 1 to 2 GHz while the frequency of the light itself is several hundred thousand gigahertz. At the detector end, the varying light signal is converted to a current signal from which the original current signal may be recovered. The next evolutionary stage in fiber optics will be the commercial availability of coherent optics where the light itself is treated as a carrier wave and amplitude, frequency, or phase modulated. When it emerges from the laboratory, this technology will provide bit rates of tens of thousands of gigahertz!

Apart from speed, fiber technology offers a number of other advantages:

- Fiber is immune to electrical interference. Because fibers are comprised of nonconductive materials, fiber is not affected by, nor does it radiate, electromagnetic energy. This can be particularly useful if you happen to live in areas where lightning is common. Besides natural radio and electromagnetic interference, fiber is also naturally immune from the man-made interference associated with motors, switchgear, and other mechanical room equipment.
- Fiber is nonconductive. Because of this, ground loop problems are nonexistent, and fiber can traverse areas where shock or explosion hazards exist.
- Fiber signal quality is far superior to copper media. Bit error rates for fiber are typically in the range of 10^{-10} or better (i.e., 1 bit in error for every 10 billion transmitted). This compares to 10^{-6} or so for copper.
- Fiber is physically small, light, and rugged.

Fiber also has some drawbacks:

- Fiber cables are relatively hard to terminate. Fiber ends require special treatment, called polishing, to make them absolutely perpendicular to the fiber's axis. In addition, special equipment is needed to splice fibers or attach connectors.
- Fiber is still not well suited to multidrop applications, such as are common in DDC. While optical splitters and Y-couplers are starting to become more common, fiber is still predominantly used point-to-point. Also, for much the same reason, fibers tend to be used to carry data only in one direction so that for a full-duplex, bi-directional connection two fibers are usually required.
- Fiber is still relatively expensive for low-bandwidth applications. At the time of writing, late 1993, typical multimode fiber cables cost about $1.00/ft compared to $0.20/ft for coax or $0.10/ft for a twisted pair.

The most common use of fiber technology in DDC systems, at least at the present time, is to provide "backbone" interbuilding communications, usually in conjunction with a standard LAN technology such as Ethernet. It is also possible to use fiber to extend low-speed communications over longer distances than would otherwise be possible. Fiber limited-distance modems are relatively inexpensive, have an EIA-232 electrical interface, and can easily transport data up to 4 km at a rate of 64 kbps. There are also fiber repeaters that have an EIA-485 interface and can therefore be used to extend or bridge EIA-485 segments over considerable distances.

6.2.3.5 The Powerline The electrical powerline can also be used to transport data. The major attraction of powerline carrier (PLC) technology is that the installation cost of interconnecting communication wiring is totally eliminated: just connect your DDC device to the AC power line and you are ready to go!

Unfortunately, there are some unique problems associated with powerlines. First, powerlines can be extremely noisy because of the effects of devices such as motors, switchgear, light dimmers, variable-speed drives, and even computers. Second, powerlines typically exhibit signal attenuation characteristics that are frequency dependent and may amount to as much as 60 dB. To make matters worse, both the noise spectrum and the attenuation characteristics can vary from moment to moment or day to day. A third problem is presented by the presence of transformers. Without some sort of active or passive bridging, PLC signals are not able to pass through.

Historically, PLC systems have been slow (< 1200 bps) and less than completely reliable due to the shifting sands of the variable powerline characteristics. Best results were obtained by measuring the noise and attenuation at the installation site and selecting a transmission frequency at the quietest, least attenuated spot. Frequencies between 50 and 500 kHz have been typically used.

More recently, at least one PLC equipment manufacturer has been able to successfully use *spread spectrum* techniques to achieve data rates up 19.2 kHz. The origins of spread spectrum communication date back to the 1950s as researchers sought to develop radio communication methods that would be less susceptible to enemy jamming. The advent of communication satellites only made the problem more critical since an enemy capable of jamming the uplink could render the entire satellite unusable. It turns out that the methods developed to deal with the military problem of intentionally generated, man-made noise can be applied just as readily to the naturally occurring noise on powerlines.

The two most common spread spectrum techniques are *direct sequencing* and *frequency hopping*. In direct sequencing systems, a single carrier frequency is first modulated by the binary data signal, then by a higher frequency repetitive pseudorandom bit pattern or code. The receiver must know exactly what the code bit pattern is and be synchronized with the transmitter. By applying the code twice, once at the transmitter and again at the receiver, the original data can be recovered. The higher frequency modulation has the effect of spreading the transmitted signal over a greater range of frequencies within a given bandwidth, thus giving the techniques its name.

In frequency hopping systems, the most common approach is to transmit sequentially on several prearranged frequencies at a rate that exceeds the data rate, effectively transmitting the same bit value several times at each of the different frequencies. If the transmitted signal at one frequency gets clobbered by a noise pulse, it may still get through at another frequency. Again, the receiver has to know when to listen on which frequency, i.e., the frequency hopping schedule has to be synchronized between the transmitter and receiver.

A second contributing factor to the improved PLC reliability that is now available is the use of forward error correction. The idea here is to include enough additional information with the data that the receiver can detect *all* errors, and correct *most* errors, without having to request a retransmission of the data. The price, of course, is an overall reduction in throughput for a given bit rate.

Although PLC undoubtedly has its place, particularly in circumstances where running dedicated cabling would be impractical or exorbitantly expensive, it is still unable to achieve the megabit per second speeds that are now commonly attainable with most LANs, even using unshielded, twisted pairs.

6.2.3.6 Physical Layer Standards: EIA-232 Electronic Industries Association Standard EIA-232, "Interface Between Data Terminal Equipment and Data Communication Equipment Employing Serial Binary Data Interchange," is one of the most prevalent data communication standards in the world and is commonly used in commercial DDC equipment. It is well worth getting to know. (Most everyone still calls the standard "RS-232" from the older nomenclature where RS stood for recommended standard, a designation that meant its implementation was voluntary. Someone, however, decided that standard numbers should be prefixed with an organization identifier, thus imposing a standard on the standards organizations!) The current, fourth version of the standard bears the suffix "D."

EIA-232 has also been adopted internationally, with minor differences, by the Consultative Committee on International Telegraphy and Telephony (CCITT), a division of the International Telecommunications Union (ITU) based in Geneva. It is one of many "V." standards (pronounced "V dot" and referred to by the CCITT as "recommendations" rather than "standards") that apply to data transmission over analog networks. Another famous group of CCITT standards are the "X." series, which refer to data communication using public data networks.

EIA-232 is actually much more than a physical layer standard, although an electrical specification is part of it. Its main purpose is to define an interface between computers (or computer terminals) and data communications equipment such as modems and line drivers, the DTE-DCE interface. The standard allows for either serial asynchronous (one character at a time with no fixed time relationship between successive characters) or synchronous communication (where the timing between sender and receiver is maintained by sending synchronization characters between data messages or special bit patterns called flags at the start of a new transmission).

The EIA-232 standard defines 20 "interchange" circuits in four categories: data, control, timing, and ground. With the exception of the protective and signal ground circuits, each circuit also has a direction associated with it: TO or FROM the DCE. The circuits most commonly found in DDC equipment interfaces, where serial asynchronous communication is used almost exclusively, are described below. (Note the absence of timing circuits. These are used only for synchronous DCEs.)

EIA-232 circuits are called *unbalanced* because the state of each circuit is measured relative to signal ground and *bipolar* because the two binary states of each circuit are signaled by voltages positive and negative with respect to ground. Circuit transmitters are required to put out ±5 to 15 VDC and receivers are expected to respond to ±3 to 25 VDC. Owing to the common availability of ±12-V power supplies in most digital electronic equipment, ±12-V is the most common voltage pair. For data circuits, −12 VDC corresponds to binary one and +12 VDC corresponds to binary zero. For timing and control circuits, −12 VDC corresponds to OFF and +12 VDC corresponds to ON.

The functions of the most common circuits are:

TD	Transmitted Data: The data from the DTE that is to be transmitted by the DCE.
RD	Received Data: Data that has been received by the DCE and is passed to the DTE for processing.
RTS	Request to Send: Control signal that tells the DCE to enter transmit mode and be prepared to transmit data. It is used primarily in half-duplex situations, i.e., two-way alternate communication.
CTS	Clear to Send: Control signal from the DCE that tells the DTE it is ready to receive data for transmission from the DTE.
DTR	Data Terminal Ready: Control signal that tells the DCE to access the communication channel. In the case of connections to switched telephone lines, this means go "off-hook."
DSR	Data Set Ready: Control signal from the DCE that indicates that the communication channel has been accessed.
RLSD (CD)	Received Line Signal Detect (commonly called "Carrier Detect"): Indicates that a DCE is receiving a "suitable" signal from the remote DCE.
SG	Signal Ground: The common voltage reference for all the other data, control, and timing signals.

Note that the standard does *not* address what happens on the line side of the DCE. Thus, even though the standard is inherently point-to-point, e.g., from computer to modem, the DCEs may be connected in a multipoint configuration, with more than two DCEs sharing the same communication line(s).

As for speed, EIA-232 specifies a rate up to 20 kbps. In practice, speeds greater than this are possible, though not too common. (Probably because the old mechanical teletype machines ran at 75 bps, speeds used for low-speed serial communication are multiples of 75, e.g., 300, 1200, 2400, 4800, 9600, and 19,200. The maximum EIA-232 speed that is often available is therefore 38.4 kbps.)

The distance limitation for EIA-232 circuits (not the DCE-to-DCE distance!) is often cited as 50 ft, or 15 m. Actually, the standard only *recommends* the use of short cables; the use of longer cables is allowed as long as the overall shunt capacitance of the cable does not exceed 2500 pF. Similarly, the load resistance, cable included, must not exceed 7000 Ω. With low capacitance cables, it is thus possible to run EIA-232 connections for a hundred feet or more.

A practical problem with the standard, and one of the reasons that books have literally been written on the subject, is that it does not require the presence of any specific set of circuits for any given application (although it does have a section on "Standard Interfaces for Selected Communication System Configurations"). It states, for example, that no data shall be transmitted unless RTS, CTS, DTR, and DSR are all on, "where implemented." For this reason, you never actually know, without reading the manual, which circuits have been implemented in a particular piece of gear. A terminal, for example, may need only TD, RD, and SG. Or it may need to see CTS in response to its own RTS. If the DCE doesn't provide CTS, the DTE can usually be tricked by setting jumpers or DIP switches (*d*ual *i*n-line *p*ackage microswitches) inside the device itself or looping the RTS output signal back to the CTS input in the connector.

If the DTE is connected directly to another DTE device, such as a computer or DDC panel, it may need a null modem to make the connection. A *null modem* is simply a cable with TD and RD switched from one end to the other (and possibly also RTS and CTS, DTR and DSR).

In spite of all its problems, EIA-232 will no doubt be with us for some time to come.

6.2.3.7 *Physical Layer Standards: EIA-485* EIA-485 is a "Standard for Electrical Characteristics of Generators and Receivers for Use in Balanced Digital Multipoint Systems." Unlike EIA-232, EIA-485 is a "pure" physical layer standard and deals strictly with the electrical aspect of bits on the wire.

Although EIA-485 provides for data rates up to 10 Mbps on a single pair of wires, its most common application in the DDC world is to provide a communication link for relatively low performance devices, such as application-specific unitary controllers, at relatively slow data rates, usually < 19.2 kbps. The reason is simply that these devices generally use inexpensive, commodity USARTs (universal synchronous-asynchronous receiver transmitters), e.g., the Intel 8251A, to drive the EIA-485 generators. USARTs are devices that convert bits from the parallel format of a computer's internal data bus to the serial format required for serial communication, and vice versa.

Because EIA-485 is "balanced," the presence of a binary zero or one is represented by the *difference* in voltage between a pair of wires, both of which are isolated from ground. A generator is required to be able to assert a differential voltage of ±1.5 to 5 VDC. If the two wires connecting a generator and receiver are labeled A and B, a binary zero is signaled when A > B by at least 0.2 VDC and a binary one when A < B by the same amount. Receivers are expected to operate over a range of Δv from −7 to +12 VDC with common mode voltages (both wires at the *same* voltage with respect to ground) of up to ±7 VDC.

The standard does not explicitly provide a distance limitation but rather supplies an algorithm for determining if a cable of known electrical characteristics will provide acceptable performance, taking into account the desired baud rate, minimum acceptable voltage level at the most distant receiver, maximum allowable bit distortion, and cable length. The cable characteristics considered are the overall DC loop resistance and risetime. For speeds in the vicinity of 9600 bps, typical twisted pairs can be run 4000 ft without difficulty.

If generator/receiver devices meet the DC loading specification of the standard, up to 32 such transceivers may be wired in parallel, or "multidropped," on a single pair of wires.

When you add the low cost of EIA-485 chips to the capabilities of speed, distance, and multiple access described above, it is easy to see why EIA-485 remains a popular standard for DDC applications.

6.3 THE DATA LINK LAYER

The primary purpose of the set of protocol procedures called the data link layer is to establish fair and equitable means by which multiple computers may access the physical communication medium. These procedures are often referred to as the *network access method.* Additionally, the data link layer is responsible for detecting, and possibly correcting, errors which may occur in the physical layer.

Data link protocols have good analogies in the protocols that govern groups of people. Imagine such a group at a town meeting. Who gets to speak? One way is to appoint a chairperson to run the meeting. It is the chairperson's job to grant speaking rights to one person at a time. Moreover, it is expected that access to the floor will be granted fairly. One way to do this might be to make sure that all who wish to speak have the opportunity to do so before anyone is allowed to speak a second time. In addition the chairperson may establish other mechanisms such as calling alternatively on speakers known to represent opposing points of view. In data communications, such a protocol, where one computer controls the access of all the others, is called a *master/slave* protocol.

We could also imagine a situation in which there is no chairperson. All the participants are then peers of one another. In this situation, someone desiring to speak could wait until a lull in the conversation, and then just speak out, hoping that no one else had the same idea at precisely the same moment. In the event of two or more simultaneous speakers, each could stop and, having counted a random number of seconds, try again. Such a procedure, implemented among computers, exemplifies a "contention" protocol.

Finally, imagine a group of Indian chiefs sitting around a council fire. Permission to speak is conveyed by passing a tobacco pipe sequentially around the circle, from one chief to the next. The rules of the council may be such that not only may the chief in possession of the pipe speak, he may also be able to ask questions, and obtain responses, during his turn. Or he may only be able to pose questions, the answers to which may only be given later when the responder to the question receives, in turn, the pipe. Such procedures are characteristic of peer-to-peer "token-passing" networks.

There are, of course, other peer-to-peer procedures that one could think of, although contention and token-passing are the most popular. Speakers could be granted a specific, recurring time slot, for example. The problem with this is that the efficiency is low if only a few people wish to speak. In between would be nothing but dead air.

To summarize, data link protocols may be classified as master/slave or peer-to-peer. Among the peer-to-peer protocols, contention and token passing are the most widely implemented.

6.3.1 Master/Slave Protocols

Master/slave (M/S) protocols have traditionally been popular in DDC equipment for two reasons: they are simple and they can be implemented with virtually any sort of physical layer technology, including multidropped modem links with EIA-232 connections to the DDC, and EIA-485 lines.

Two approaches to M/S are common. The first is called *request/response* and the second is called *poll/poll-response.* In request/response systems, the slave is only allowed to respond to the specific request or command previously directed to it. The slave has no opportunity to initiate any communication. Typical requests would be to send back the status or value of specific hardware or software points, to output new values to a binary or analog control point, or to initiate some specific sequence of control operations. The responses to these requests could contain the desired data or simply an acknowledgment that the requested control action had been attempted or actually taken.

In poll/poll-response systems, the slave is sent a special message called a poll, which may be as short as a single byte containing the numerical address of the slave. Upon receipt of a poll, the slave may send any message it desires to the master. Such messages include notifications of changes of state of monitored or controlled points, alarms, and various types of error messages, including requests for software downloads.

Most commercial DDC systems of the 1980s used the M/S network access method and, more often than not, provided both request/response and poll/poll-response capabilities. Note that the use of M/S data link protocols does not necessarily preclude peer-to-peer communication. The master may be programmed to pass along information from one slave to another. Suppose we want to share the value of a particular sensor, e.g., the outside air temperature. When the master detects that a change-of-value has occurred, either by reading the value of the point or receiving a change of value message in response to a poll, it can retransmit the data to specific recipients or, perhaps, to all slaves simultaneously via a “broadcast” message.

6.3.2 Peer-to-Peer Protocols: Contention

Increasingly, local area network technology is being used by DDC manufacturers. One of the oldest, most widespread of these is Ethernet, the subject of Section 6.3.7. Ethernet uses the contention network access method.

The most common contention protocol is called *carrier sense multiple access with collision detection* (CSMA/CD). A station wishing to transmit first listens to the medium to see if it is in use or not, a process called *carrier sensing.* If the channel is clear, the station begins its transmission and simultaneously copies whatever it hears on the medium into its receive buffer. If the data in the receive buffer does not agree bit for bit with the message it is outputting, the station concludes that another station must have begun transmitting at, or nearly at, the same time. This process of looking for garbled data is called *collision detection.* The CSMA/CD procedure then calls for the station to continue transmitting for at least as long as the *jam time,* a period long enough to ensure that all stations on the network are aware of the collision (which they can detect by noting that the message does not contain valid error-checking information). The protocol then requires the colliding stations to "back off," i.e., cease transmitting, for a time determined by multiplying a random number by the minimum message length time. After deferring for this random length of time, the stations are free to try again.

Although it may seem that networks based on CSMA/CD are somewhat hit and miss, which they are, studies have shown that the throughput of such networks usually exceeds that of the more deterministic token-passing networks—up until the time the network loading approaches 70% or thereabouts. This is just the result of the fact that no overhead is consumed by passing the token message to stations with nothing to say. As with any network, the success of the CSMA/CD approach requires studying the potential network load and adjusting the number of stations on any particular network segment as necessary to achieve the desired throughput a statistically reasonable percentage of the time.

6.3.3 Peer-to-Peer Protocols: Token-Passing

In token-passing (T/P) networks, the right to use the physical medium is conveyed from one station to the next by means of a special message called the *token.* Exactly what rights are conferred by possession of the token varies from one specific type of T/P network to the next. In some T/P networks, the station with the token appends its data to the token, which is then sent to the next station in line. When the station to which the message was actually directed receives the token, and the accompanying message, it strips off the message and replaces it with its response, which then eventually makes its way back to the original sender. This type of operation is characteristic of token rings.

In the case of token buses, where all stations hear all traffic, possession of the token may allow the possessing station to send messages to, and receive messages from, other stations during its time of possession without having to wait for its respondent station to receive the token. This technique amounts to sequential M/S operation: a station gets to be the master for as long as it is legally allowed to hold the token. Moreover, this type of T/P network allows the mixing of stations with different capabilities on the same medium. Some stations may never receive the token, acting only as slaves when spoken to.

In comparison with CSMA/CD networks, T/P networks have one arguably significant advantage: they are "deterministic." This means that it is possible to ascertain, with reasonable certainty, the amount of time it will take for any particular station to be able to gain access to the network. It is for this reason that T/P networks have found favor in time-critical applications such as factory automation and process control. One T/P network, ARCNET, is gaining popularity in DDC applications and will be discussed in Section 6.3.6.

On the other hand, T/P networks have some drawbacks, the greatest of which is that they are the most complicated to implement due to all of the possible error conditions that can arise. Not surprisingly, the most onerous problem is the loss of the token. This can occur if the token message is garbled due to noise on the transmission medium or if a station, having properly received the token, dies before it can pass the token to the next intended recipient. Whatever the circumstances, the network has to be able to regenerate the token and, possibly, rediscover which stations are currently on line. The logic used by ARCNET will serve to illustrate one way this is done.

6.3.4 Network Topologies

No discussion of networks can be considered complete without at least some discussion of network *topologies*. The topology of a network describes the geometric layout of the wiring between participating nodes. Each topology has pros and cons.

The three fundamental arrangements are the bus, the star, and the ring. See Fig. 6.5. Other arrangements can be made by combining the three basic types. The bus is no doubt the simplest: cable is simply run from one node to the next and the node is connected by a T-splice or tap of some kind. Often the ends of the bus have to be terminated by a resistor of the cable's characteristic impedance to prevent signal reflections due to an impedance mismatch. One of the principal characteristics of a bus is that all nodes hear all the transmissions on the link. The down side of this is that a shorted cable, or a node that has gone "nuts" and is transmitting constantly, can take out the entire bus.

A star consists of a hub and a series of radial connections, or spokes, to each node. The hub may be either electronically active or passive. Active hubs usually perform signal amplification between the spokes and may even be smart enough to respond to management commands, allowing a remotely situated network manager to collect information on error conditions that the hub has accumulated or to cause the hub to electrically isolate a spoke which is causing problems. Passive hubs simply interconnect the spokes and may contain some components to provide impedance matching.

The major advantage of the star topology is that all the spokes come together at the hub so that network monitoring and troubleshooting can often be done without leaving the wiring closet containing the hub. The disadvantages are the single point of failure represented by the hub and the fact that hub-spoke wiring generally requires running far more lineal feet of cable than the bus topology. Stars are particularly popular in office automation situations where there are sometimes already extra cable pairs at each telephone jack that lead back to the central building wiring closet. In these cases, the use of star hubs allows these existing pairs to be pressed into service for data communications without further labor.

The final topology is the ring, a topology whose network access method is always token-passing. In a true ring, each node is hardwired to its neighbors on either side until the ring is completed. Unlike nodes connected via bus or star topologies, each node only hears transmissions from its upstream neighbor. Because of this, the

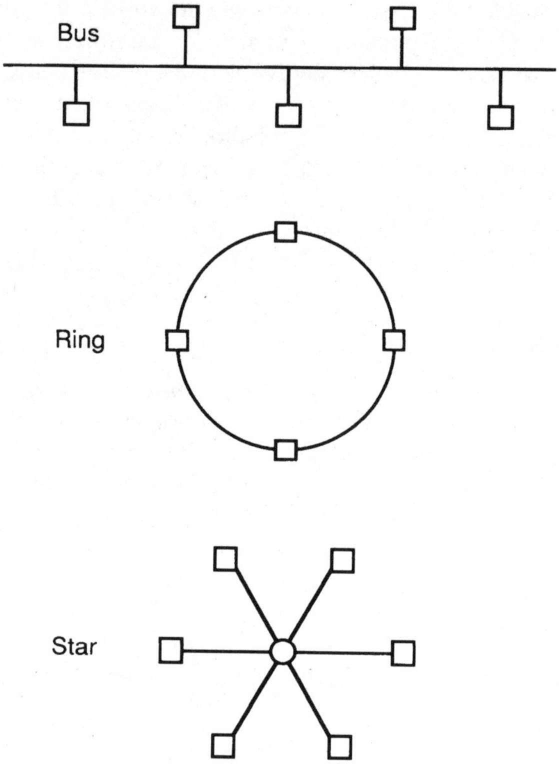

FIGURE 6.5. The three most common network topologies. Depending on the specific network technology, the hub at the center of the star may be electrically passive, simply interconnecting the radials, or active, containing amplifiers and, perhaps, network management software.

procedures for token regeneration are the most complicated of any of the token-passing networks.

This is a good place to point out the difference between a *physical* ring and a *logical* ring. A physical token-passing ring is as just described. A logical token-passing ring, which may be implemented on either a bus or a star, consists of a list of stations which pass the token from one to the next as if they were on a ring, from the first station on the list to the second, from the second to the third, and so on, until, finally, the last station on the list passes the token back to the first.

Several ways have been developed to make physical rings more manageable. The first is that the wiring to and from each node can be routed through a central location creating, in effect, a "ring-star" (not to be confused with Ringo Starr, the erstwhile drummer of the Beatles in the sixties). The ring remains electrically a ring but the wiring becomes accessible at one point, a great aid to problem isolation, since a station believed to be causing problems can be jumpered out at the "hub." The IBM 4- and 16-Mbps token ring uses this approach, the hubs being referred to as *multi-*

station access units (MAUs).

The second practice is to run redundant cabling so that actually two rings are formed. The second ring may only be used for standby purposes or a separate, independent token may circulate on it. If the token circulates in the opposite direction to the first, the rings are called *counterrotating.*

A practical network that uses dual counterrotating token rings is the 100-Mbps Fiber Distributed Data Interface (FDDI) that has been developed by ANSI committee X3T9. FDDI supports up to 500 stations with an overall ring circumference of up to 200 km. At present, however, the capabilities of the technology, not to mention the cost of FDDI hardware, are far in excess of what is typically required for DDC.

6.3.5 LAN Standards: IEEE 802.X

Recognizing the role standards usually play in promoting specific technologies, the Institute of Electrical and Electronics Engineers (IEEE) formed a project committee in 1980 to establish networking standards for the burgeoning LAN market. The project was called Project 802—Local and Metropolitan Area Network Standards. The committee had its work cut out for it! After considerable debate and strife, two conclusions were quickly reached. First, the LAN problem was too complex to be considered as a single monolithic problem; it would have to be broken up into a number of more manageable subproblems (the same conclusion reached by the ISO). Second, no single technology could satisfy all the requirements that the committee brought to the table.

These conclusions found expression in the development of a new reference model along the lines of the ISO's model. It consists of the bottom two layers of the ISO stack, the data link and physical layers, but with the data link layer divided into two sublayers. The first is called the logical link control (LLC) sublayer and the second is called the medium access control (MAC) sublayer. Beneath the LLC and MAC sublayers is a physical layer appropriate to the medium in question. See Fig. 6.6. This model turned out to provide a framework capable of accommodating all of the diverse points of view of the committee. Prominent among these was a preference for either bus or ring topology; within the bus topology, for either CSMA/CD or token-passing, and either baseband or broadband physical layer signaling.

To deal with this matrix of technologies, the committee was organized into 10 subcommittees:

- 802.1: High Level Interface Issues
- 802.2: Logical Link Control
- 802.3: CSMA/CD Networks
- 802.4: Token Bus Networks
- 802.5: Token Ring Networks
- 802.6: Metropolitan Area Networks
- 802.7: Broadband Issues

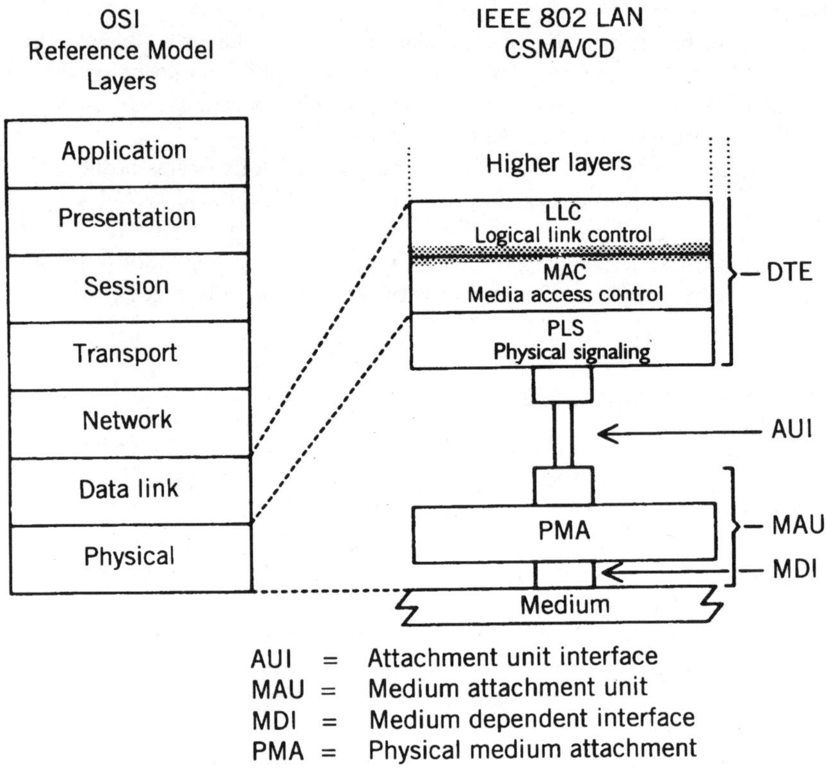

FIGURE 6.6. The relationship between the ISO OSI model and the IEEE 802 series LANs. (Reprinted from ANSI/IEEE Std 802.3, IEEE Standard "Carrier sense multiple access with collision detection (CSMA/CD) access method and physical layer specifications," copyright © 1992 by the Institute of Electrical and Electronics Engineers, Inc. with the permission of the IEEE.)

- 802.8: Fiber Optic Issues
- 802.9: Integrated Voice and Data Networks
- 802.10: Data Security and Privacy Issues

The task of each committee has been either to develop or adopt appropriate standards in its area of concern. Given this choice, several of the committees have chosen to adopt existing LAN technologies with little or no modification. The 802.3 committee, for example, adopted Ethernet, a LAN we will discuss in more detail shortly. Similarly, the 802.5 committee has put its stamp of approval on IBM's 4/16 Mbps token ring.

Of the work accomplished by the committee to date, that of the first three subcommittees is most likely to find its way into DDC, at least in the near term. The work of 802.1 is important because it sets out the framework within which the other standards have been, or are being, defined. 802.2 LLC is important because it defines the protocol services that the entire family of 802 standards provides. 802.3 CSMA/CD is important because several DDC vendors are currently offering products which conform to it, and it is one of the LANs adopted by the ASHRAE BACnet committee.

Since 802.2 LLC is the highest layer in all of the 802 family of LANs, and is also used by ANSI's FDDI, it is worth a few words. Its function is to provide the protocol means to exchange data between peer machines, regardless of the underlying MAC technology. It does this by defining three forms of service which differ in reliability and efficiency and may be therefore more or less suited to specific applications. The first type of service is called *unacknowledged connectionless* service. In this form of LLC, messages are simply addressed to a destination and *transmitted*. There is no sequencing of messages or acknowledgment that the message arrived.

At this point it is vital to distinguish between data link acknowledgments and acknowledgments provided by higher-level protocols or programs. Suppose, for example, that a DDC program needs to find out a temperature somewhere and the device in which the program runs uses unacknowledged connectionless LLC. The information request is formatted, in the appropriate higher-level protocol format, and sent out. The fact that no data link acknowledgment is received does not mean that the communication is inherently hit-and-miss. The responsibility for determining what to do if no response comes back is simply assumed by the original requester. If no answer is received, it is up to the requesting program to decide what to do. The possibilities depend on the program. The first thing to do would be to retry the transmission, possibly two or three times. If still no answer is received, the program would usually notify its user, either by displaying something on the screen, writing something to a database, or taking the control action appropriate to a loss of communications.

A somewhat more "robust" form of LLC is called *acknowledged connectionless*. In this case, an explicit acknowledgment is sent to the originating data link for each message correctly received. Notice, however, that a data link acknowledgment cannot take the place of an acknowledgment from the destination application program. The data link ACK merely indicates that the destination machine is alive and well, not that the desired program or process in the destination machine is accessible. Nonetheless, acknowledged connectionless service may be useful in critical applications, particularly where several LAN segments are interconnected by nodes charged with the responsibility of routing messages from one physical data link to another, such nodes being usually called *routers*. In this case, where a message is received on one LAN and retransmitted on another LAN to reach its ultimate destination, there arguably may be value in finding out, more or less immediately, that the message at least reached, or did not reach, the router.

The third, and most sophisticated, form of LLC is called *connection-mode* service. In connection-mode service, a logical "connection" must be made between two machines prior to the exchange of data. Once established, a connection provides certain benefits including automatic message sequencing and message flow control.

Nonetheless, connection-mode service is rarely, if ever, used in DDC or other real-time process control applications. Among the reasons are the time burden of having to establish a connection before being able to exchange useful data, the requirement to establish a separate connection with each and every communication partner, and the amount of computing resources the foregoing would require.

In the end, the simplicity of unacknowledged connectionless LLC service seems, for DDC communications at least, to be the most suitable.

6.3.6 LAN Standards: ARCNET

ARCNET has the distinction of being the first commercially available LAN in the world. It was introduced in 1977 by the Datapoint Corporation as a means of connecting "attached resource computers," their designation for an early form of desktop computer, to shared printers and file servers. As of 1991, there were reputed to be in excess of 2 million ARCNET nodes. The primary reason is that ARCNET is reliable, time-tested technology—and the least expensive LAN available operating in the > 2-Mbps range. These are also the factors that have favored the introduction of ARCNET into the DDC world.

But ARCNET has had its problems. First, it is having to shed its image as a tightly controlled, proprietary technology. Even now, the ARCNET chips that are at the heart of each commercial product are only available from two sources. Second, many potential LAN purchasers are put off by the relatively modest signaling rate of only 2.5 Mbps (as compared with 4 and 16 for 802.5 token ring and 10 for 802.3/Ethernet). A final objection, that for years ARCNET was only a *de facto* standard, was put to rest in October of 1992 when ARCNET was officially blessed as an ANSI standard, ANSI/ATA 878.1 (ATA is the ARCNET Trade Association).

Nevertheless, over 70 manufacturers make various types of ARCNET products, both for personal computers and the original equipment manufacturer (OEM) market. And to address the concern about speed, a 20-Mbps version of ARCNET has been announced, although it is not yet actually available (a phenomenon cynics refer to as marketing "vaporware!").

What is ARCNET? It is actually a LAN specification for the lowest two OSI protocol layers, the physical and data link. As such, it specifies electrical signaling, media characteristics, and a protocol for accessing the network media. Let's look at each of these aspects of ARCNET in more detail.

At the physical layer, ARCNET represents zeros and ones by "symbols," each of which is exactly 400 ns in duration. A zero is represented by the absence of voltage for this period while a one is represented by one complete voltage cycle of a sine wave lasting 200 ns followed by 200 ns of silence. The bit rate is the reciprocal of the bit period = $1/(400 \times 10^{-9})$ = 2.5 Mbps. The peak-to-peak voltage at the transmitter is supposed to be at least 15.4 V. Each 8-bit data octet is preceded by three symbols, B′110′, which provide synchronization and a rudimentary error detection capability.

ARCNET can use coax, unshielded twisted pairs, or fiber. The coax has to be RG-62, which has a characteristic impedance of 93 Ω. The specification calls for the coax to be terminated with BNC connectors. For twisted pairs, standard telephone-type RJ-11 modular plugs are recommended. The use of fiber, while recognized in the proposed ATA standard and commercially available, is not further detailed in the specification.

The topology of ARCNET can be either bus, star, or a combination of the two. Since there is also no restriction on the intermingling of various media, ARCNET LANs can be configured with great flexibility. The key is the use of hubs, which may be either active or passive. Using coax, nodes may be as far as 2000 ft from an active hub and up to 10 active hubs may be daisy-chained, giving an overall network extent of 22,000 ft, about 4 miles. Technically, this distance is the consequence of stipulating a maximum allowable propagation delay of 31 μs and allowing each hub to take up to 700 ns to rebroadcast the incoming signal. (Although I have no proof, I suspect the designers backed their way into the 31-μs value by deciding they wanted to build a 4-mile wide LAN!) In any case, greater distances can be achieved with fiber, and with twisted pairs the distance between hubs is cut down to 400 ft.

At the data link layer, ARCNET is a logical token-ring, implemented with five extremely succinct message types. This brevity is at once a virtue and a liability. Node addresses, for example, are limited to one octet. Therefore, the maximum number of nodes that can exist in one ARCNET domain is 255, the address 0 being reserved for broadcast messages. Also, the maximum number of data octets per transmission is only 507. As a result, ARCNET thrives where the number of nodes is limited and short messages are the rule, a situation that is often found in the DDC environment. (As we will see in the next chapter, it *is* possible to extend the number of interconnected ARCNET nodes, but the maximum message size of standard ARCNET is truly fixed at 507.)

ARCNET Plus, the 20-Mbps version of ARCNET, will substantially alleviate the two restrictions of address space and message length, allowing 2047 nodes and 4000 octet messages, respectively. Another major, and somewhat amazing, feature of ARCNET Plus is that it is supposed to be able to coexist with 2.5-Mbps nodes *on the same physical medium*, although, not surprisingly, any existing hubs will have to be upgraded!

Let's look at how the token-passing medium access control part of ARCNET actually works. First, there are five message formats, also called frame types, that are used for protocol control and data transfer. They are: (1) Invitation to Transmit (ITT); (2) Data Packet (PAC); (3) Free Buffer Enquiry (FBE); (4) Positive Acknowledgment (ACK); and (5) Negative Acknowledgment (NAK). There are also nine different timers, eight internal binary "flags," and four internal 8-bit registers. All the timers, flags, and registers are implemented in a single integrated circuit. Fortunately, we don't really need to understand the functioning of all these components to get a reasonable feel for the protocol.

The ITT message is actually the token. It is passed sequentially from one station (the official ARCNET terminology) to the next in ascending order of address, with the station with the highest address passing the ITT to the member of the ring with the lowest address. Each station knows its own address and the address of the station to which it should pass the ITT token. These two addresses are stored in two of the registers mentioned above.

In the case where no station has any data to send, the token is simply sent from one station to the next. The sending station considers that the token has been successfully passed if it hears a binary one, i.e., activity, before one of its nine timers, the "activity timeout" timer, expires. If, however, this timer does expire, as would occur if the next station in line "died" for some reason, the token-passing station then starts the reconfiguration procedure. For this case, the procedure is very simple. The station simply increments its stored address of the "next" station and tries to send the ITT to it. This goes on until the "next" station is located, whereupon normal operation resumes.

If a station has data to send when it receives the ITT, it either sends it immediately, if it is a broadcast message, or sends an FBE message to the specific desired recipient. If the recipient has an available free buffer, it sends an ACK back to the token holder. The token holder then sends its data in a PAC message. If the recipient receives the message correctly, based on the validity of the "frame check sequence" error detection octets at the end of the message, it again sends an ACK to the token holder who then passes the token up the line. The data portion of the PAC message contains the DDC-related information. Suppose it is a request to send back a particular temperature reading. The receiving station waits until it receives an ITT and then responds by going through the sequence of FBE/ACK/PAC/ACK just described in the reverse direction. If either the FBE or the PAC is NAKed, or no response at all is received, the token holder is required to immediately pass the token to the next station in line. What recovery action is to be taken is determined by the higher-level DDC protocol.

What happens if a new station wants to join the ring? The answer, at first, sounds horrendous. It sends out a blast of 765 9-bit symbol groups called a *reconfiguration burst*. The length of the burst, about 3 ms, is designed so that it forces the token to be destroyed and all other activity on the network to cease. Each station then enters its reconfiguration mode. This involves starting a special timer whose duration is inversely proportional to the station's own address. Each station then keeps watch on this timer and listens for activity on the bus. Eventually, the station with the highest address times out and tries to pass the token. All the other stations, each with lower addresses, then breathe a sigh of relief and enter the state where they each start to listen for their own address as the next possible token recipient.

Since each station is only listening for *activity* and not for its own address while waiting to find out if it is the highest numbered station on the bus, the first message sent by a station when it times out is an attempt to pass the token to itself! Since that inevitably fails, it increments the value of the next station register (which had been set as part of the reconfiguration procedure to its own address) and tries again. Eventually, the next station in line is found and the token is considered passed. This process repeats until each station on the bus knows who its next door, upstream neighbor is.

The identical process, reconfiguration burst and all, is carried out if the token gets "lost." Although this is a rare event (hopefully), it can occur if, after receiving the token, the token holder dies. While the reconfiguraton logic may seem quite involved, and the length of time it takes to accomplish it depends on the number of stations and what their addresses are, the good news is that it can generally be accomplished in less than 60 ms!

In summary, ARCNET provides a reliable, speedy, and inexpensive technology that will probably be around in DDC LANs for some time to come.

6.3.7 LAN Standards: Ethernet/802.3

While ARCNET may be able to claim the most installed nodes, at least into the 1990s, the best known of all LANs is certainly Ethernet and its derivatives standardized by the IEEE 802 committee and the ISO. Ethernet is the archetypal contention, CSMA/CD LAN. Owing to its widespread popularity, declining cost, and recent adaptation for use on unshielded, twisted pairs, it is finding increasing use in DDC LANs.

Ethernet, like ARCNET, is both a data link and physical layer specification. It was first developed by Xerox in 1976 and patented in 1977. The first version ran at 3 Mbps using baseband signaling on coaxial cable. In an effort to promote the technology, Xerox joined forces with Digital Equipment Corporation and Intel. Together they developed specifications and hardware. The Version 1.0 specification appeared in September, 1980, and two years later, Version 2.0 was published. The main difference between the two versions was an enhanced physical layer specification dealing with self-testing of the contention detection circuitry and the automatic squelching of transmissions recognized to be too long, the so-called *jabber* function.

More or less concurrently, the 802.3 subcommittee was drafting a CSMA/CD standard based on the Ethernet technology which would embrace the LLC functionality proposed by 802.2. The standard was first published in 1985 as ANSI/IEEE 802.3-1985 and ISO Draft International Standard 8802/3. Since then, a revised, full-fledged ISO version has appeared as 8802-3:1989, and the IEEE has come out with five supplements that extend the use of the standard to media other than the original coax. The standard was revised again in 1993.

To understand this flurry of activity, let's back up and look at the original Ethernet in more detail. It was intended to run on only one, very specific type of 50-Ω coaxial cable. The cable was 0.405 in. in outside diameter and relatively stiff and rugged. Because of this, thick-wire coax, as it is now called, could not be easily brought to the computer. Rather, the concept was that the computer would have to be connected to the coax by means of some kind of interface cable. The device physically attached to the cable was called a transceiver by Xerox, et al., and later a *medium attachment unit* (MAU) by the IEEE. The computer-to-MAU cable itself, originally called simply a transceiver cable, is now referred to as an *attachment unit interface* (AUI) cable. Ah, standards committees!

In any event, the IEEE committee soon began to realize that there was a need to be able to standardize media other than the original, baseband thick-wire coax. The idea was that the part of the physical layer residing in the DTE that actually generates and receives the data could then remain fixed, and you would merely have to plug your AUI cable into a MAU appropriate to the new medium. In order to minimize confusion, the committee came up with a naming convention to designate the new physical medium specifications that were under development. The designation has three parts:

<data rate> <type of signaling> <maximum segment length>

where the data rate is in Mbps, the type of signaling is either "Base" for baseband or "Broad" for broadband, and the maximum segment length is in hundreds of meters. In those cases where the maximum segment length is still under discussion, a letter designation for the type of medium is used. Thus the original Ethernet and 802.3 specification describes the 10Base5 medium: baseband signaling at 10 Mbps for up to 500 m.

With this shorthand, we can now indicate, at least in part, the subject matter of the 802.3 supplements. They describe 10Base2, 10Broad36, 1Base5, and 10BaseT where "T" stands for twisted pair. The 10BaseF fiber optic specification is still under development.

10Base2 is the 802.3 variant designed to operate over thin-wire RG-58 cable. It has been dubbed by some "Cheapernet." The main advantage of the thin coax, which is about half as thick and vastly more flexible than the original thick-wire Ethernet coax, is that it is much easier to work with and is far less costly. Although thin-wire MAUs are available, it is possible to obtain network interface cards that are self-contained and attach directly to the RG-58 by means of a BNC T-connector. The main limitation is distance. As the 10Base2 designation indicates, the maximum segment length can only be 200 m (actually 185 m rounded off). Still, this is almost the length of two football fields. With repeaters, 10Base2 segments can be interconnected for an overall network span of 925 m. Thin-wire segments may also be attached to thick-wire or fiber optic backbones.

10Broad36 also uses coax but of the 75-Ω cable TV variety. Broadband systems require a "head-end" repeater which takes in transmissions and rebroadcasts them, either on a separate cable or in a different frequency band. The first configuration is called dual-cable while the second is called split-frequency. The great allure of broadband systems is that other services such as TV or digitized telephone can share the same cable simply by using different frequency bands. To reduce the required bandwidth, broadband MAUs convert the Manchester-encoded data to NRZ format (remember that Manchester encoding requires roughly twice the bandwidth of NRZ since there is a 0 → 1 or 1 → 0 transition during every bit), scramble it according to rules given in the standard, then use the scrambled data to modulate an RF carrier using differential phase shift keying. The purpose of the scrambling is to try and break up any long strings of zeros or ones that are common in many types of data so that timing information may more easily be extracted at the destination, one of the advantages of the original Manchester-encoded data stream. At the destination the process is reversed to recover the original information.

The 1Base5 physical specification is often called "StarLAN" and was introduced by AT&T. Unlike the previous technologies, 1Base5 was designed to operate on unshielded, twisted pairs (UTP) in a star rather than bus topology. The network management advantages of the star topology have already been presented. With the approval of the 10BaseT specification in 1990, it is unlikely that 1Base5 will see much further growth.

10BaseT (802.3i) is full 10 Mbps 802.3 over UTP. Like 1Base5 it is also designed to operate with hubs. Because the data rate is identical to that of the coax versions, 10BaseT hubs can easily be bridged to other 10 Mbps LAN segments. The distance limitation is 100 m from DTE to hub. 10BaseT requires two pairs, one for transmitting and one for receiving. The connectors are standard 8-conductor RJ-45s (ISO 8877). A number of challenging problems in the areas of frequency-dependent attenuation, crosstalk between adjacent pairs, collision detecton, and MAU reliability had to be solved in order to make 10BaseT work. We will return to these shortly.

Finally, fiber optics is beginning to appear in 802.3 networks. This is the subject of the 10BaseF specification which is still in the works. Nonetheless, *fiber optic inter-repeater links* (FOIRL) are commercially available now and permit the interconnected LAN segments to be located as far as 10 km from each other.

How does the 802.3 medium access control protocol work? The same basic algorithm is used for all media:

1. Listen to determine if the medium is idle and, if so, transmit. Otherwise go to 2.
2. If the medium is in use, continue to listen until the medium is determined to be idle, then transmit immediately.
3. If a collision is detected, transmit a jamming signal of sufficient duration that all stations on the LAN can detect that a collision has occurred. Then cease transmitting.
4. After a random interval, go to 1 and try again (up to 16 times).

See Fig. 6.7. Several timing issues are raised by this algorithm. For example, what is the maximum amount of time it might take to detect that a collision has occurred? To figure out the answer, imagine a worst-case scenario wherein the two stations at the farthest extremes of the network start transmitting simultaneously. Neither station will be aware of a collision until the signal from the remote station has reached it. This time, the time it takes for a signal from one end of the network to propagate to the other, is called the *slot time.*

The slot time is crucial to the operation of CSMA/CD. Not only is it an upper bound on how long it takes to detect a collision or, alternatively, how long a station must wait to be assured that it has acquired uncontested access to the medium, it also sets a *lower bound* on how long a message must be in order that all stations may detect that a collision has occurred. The reason for this has to do with how collisions are detected. In bus configurations, a collision is signaled if the MAU receiver detects a voltage at its input greater than its own transmitter can put out by itself. In star systems, a collision is deemed to have occurred if a hub senses activity on more than one input at the same time. In broadband systems, the MAU does a bit-by-bit

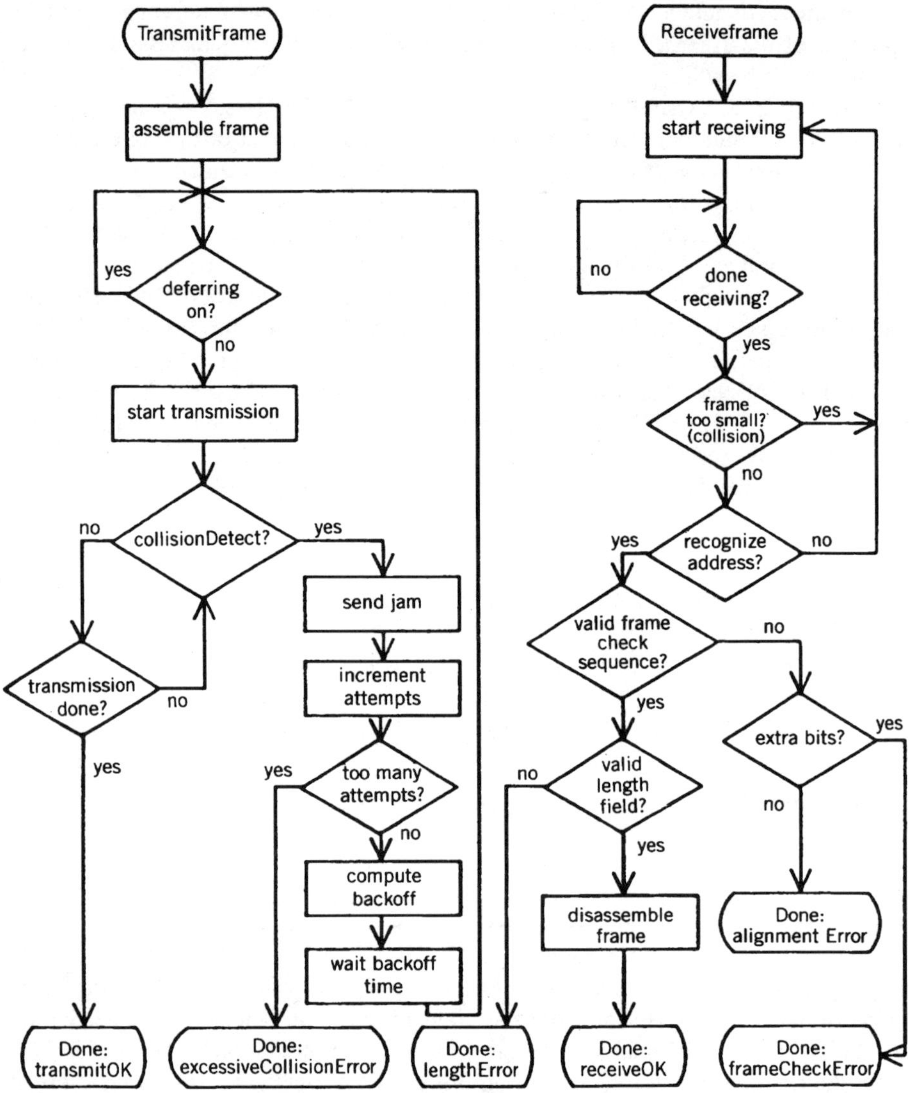

FIGURE 6.7. The basic CSMA/CD algorithm used by the Ethernet/802.3 LAN. (Reprinted from ANSI/IEEE Std 802.3, IEEE Standard "Carrier sense multiple access with collision detection (CSMA/CD) access method and physical layer specification," copyright © 1992 by the Institute of Electrical and Electronics Engineers, Inc. with the permission of the IEEE.)

comparison of the signal it is sending with the rebroadcasted version of its own transmission from the head end. In *all* of these cases, the transmission must be going on at the time of collision detection.

Given a 2.5-km network with the maximum number of repeaters and some assumptions about expected delays within them, the 802.3 specification has come up with a slot time of about 58-μs or about the time it takes to transmit 580 bits at 10 Mbps. This value determines the minimum packet length: after stripping off the preamble, starting frame delimiter, and frame check sequence octets, this leaves 60 octets for addresses, length field, LLC information and data. Packets with less data are padded with extra bits to reach the minimum size needed for reliable collision detection. The maximum packet length is somewhat arbitrarily set at 1526. After stripping off all overhead, this leaves exactly 1500 octets for LLC control and data.

Once a collision is detected, it has to be "enforced" so that all stations know about it. This is the purpose of the jamming procedure on bus and star networks. The 10Broad36 specification allocates 14 Mhz for data and another 4 Mhz for use by any MAU to transmit a signal to all other stations that it has detected a collision.

One of the worst things that can happen in a CSMA/CD network is for a hardware failure to occur whereby a station does not get valid collision detection information. In this case, the station may just transmit away, unaware that its messages are colliding. Without collision detection, the network can be easily crippled. To protect against this possibility, Ethernet V2.0 and 802.3 incorporate the idea of a "heartbeat" signal from the MAU to the DTE. This amounts to a periodic test of the collision detection circuitry. If the DTE does not get a heartbeat message from the MAU on schedule, it figures that a hardware failure has occurred and notifies its operating system accordingly.

In the case of 10BaseT, a failure of the receiving pair could cause the same problem. To protect against this, each MAU sends a link test pulse approximately every 16 ms. The pulse is brief enough so that it is not interpreted as the start of a valid frame but gives each receiver an opportunity to verify that it is still connected to a MAU at the other end.

The 802.3 packet format is shown in Fig. 6.8. Each packet begins with a 7-octet preamble of X′55′ or B′01010101′. This pattern of zeros and ones assures a nice square wave for use by the receiver to synchronize its clocking circuits. These are followed by a *start frame delimiter* (SFD) of X′D5′. Since the least significant bit is transmitted first, the SFD ends the sequence of alternating zeros and ones with two ones back to back.

Although MAC layer addresses of either 2 or 6 octets are allowed, the 2-octet format is rarely, if ever, used. Using 48 bits, the number of addresses available is $2^{48} \approx 3 \times 10^{14}$, enough so that each 802.3 device in the world (universe!) can have its own, unique address. To ensure that no two vendors happened to pick the same addresses for their products, Xerox and later the IEEE assigned manufacturers values for the first three octets, the so-called organizationally unique identifier. Vendors could then assign the last three octets themselves with the assurance that the addresses so selected would be guaranteed to be unique.

Following the source and destination addresses is a 2-octet length field, followed

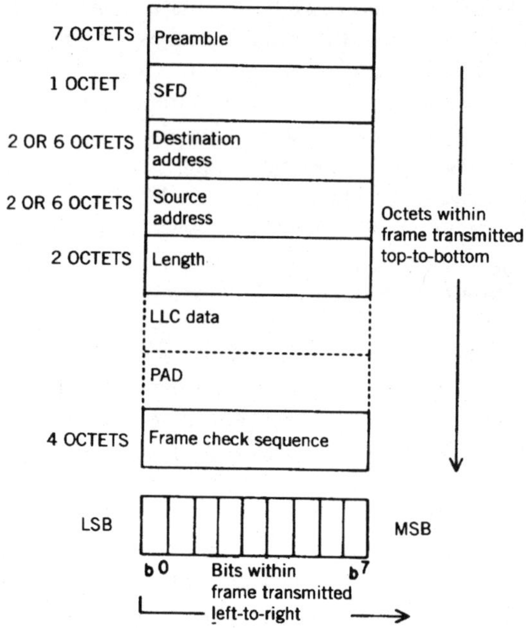

FIGURE 6.8. The format of an Ethernet/802.3 message. (Reprinted from ANSI/IEEE Std 802.3, IEEE Standard "Carrier sense multiple access with collision detection (CSMA/CD) access method and physical layer specification," copyright © 1992 by the Institute of Electrical and Electronics Engineers, Inc. with the permission of the IEEE.)

in turn by the LLC data. In the case of unacknowledged connectionless frames, this amounts to 3 octets of link address and control information followed by up to 1497 octets of data. A valid frame ends with a 4-octet *frame check sequence* that is used by the receiver to verify that the frame was actually received without error.

Given the time-tested reliability, declining cost, and mixed media capability of Ethernet and its derivatives, it is clear why an increasing number of DDC manufacturers are selecting it for their high-performance applications.

6.3.8 Data Link Error Detection

An important function of the data link layer is to detect errors that may occur in the course of a message making its way through the network. Errors may result from signal attenuation, various types of distortion, electrical noise, or, in the case of CSMA/CD networks, collisions. The importance of error detection varies. In digital telephone systems, for example, the human ear can compensate for some amount of garbling and still understand the voice at the other end. Direct computer-to-computer communications are usually not as forgiving, and in the case of DDC, a garbled message could potentially cause a critical building system to malfunction or an erroneous alarm to be generated. This may be taken as further proof, if any is needed, of the sophistication of the computer known as the human brain!

Although the science of error detection in its most academic form is extremely complex, the underlying idea is simple: additional information is sent along with the original message that can be used by the receiver to determine that the message has actually been correctly received. The form and amount of the "additional information" varies greatly as does its ability to detect errors. The effectiveness of different error-detecting methods is calculated using probability theory. The most sophisticated methods can claim that the odds of an undetected bit error slipping through are better than one in a billion.

One of the simplest error detection methods makes use of the property of bit strings known as *parity*. A string of bits is said to be of even parity if the total number of 1 bits present is an even number, e.g., 0, 2, 4, 6, or 8. If the number of 1 bits present is an odd number, e.g., 1, 3, 4, 7, or 9, the string is said to be of odd parity. In mark or space parity, the parity bit in a given string is always 1 or 0, respectively. Parity checking alone can only detect a very limited range of errors, namely, those where an odd number of bits have been altered by transmission interference in the case of even or odd parity checking or where the parity bit itself has been affected in the case of mark or space parity checking.

Parity checking is often used with the 7-bit ASCII code in serial, asynchronous transmission. In the case of ARCNET, the sequence 110 precedes each group of eight data bits. If the receiver does not see the 110 sequence, it knows immediately that an error is present. This is similar to the use of mark or space parity using three bits instead of one.

Another obvious, though not very efficient technique is simply to send the message twice. ARCNET does this in the case of the destination address field. Since ARCNET addresses are only one octet long, the overhead is minimal.

If the binary values of each message octet are added together, the result can be used for error checking. The value thus obtained is called a *checksum* and can be appended, in whole or in part, to the original message. The receiver then performs the same addition and compares the result against the received checksum to verify that they match.

The most powerful, and increasingly common, technique uses what is called a *cyclical redundancy code* (CRC). This code, which is also appended to the transmitted message, is generated by subjecting the message data stream to a specific set of mathematical operations that are performed in the identical manner by the receiver. If the codes match, the message is considered to have been correctly received. The most common operation is to divide the transmitted data stream by a specific n-bit binary number that has been found mathematically to possess desirable error-detecting properties. The remainder after division is used as the CRC. Because the divisor is a binary number, the remainder will always have one bit less. Thus, if it is desired to have a 16-bit CRC, a 17-bit divisor is used.

Virtually any degree of reliability can be attained by increasing the number of bits in the error code. The frame check sequence mentioned in the discussion of 802.3 is a 4-octet value derived from a standardized 33-bit divisor. The derived CRC is known as CRC-32. Apart from their effectiveness in detecting errors, CRCs are popular because the mathematical operation of binary division can readily be implemented with shift registers and exclusive-OR gates, thus allowing CRC processing to be performed by hardware quickly and cheaply.

6.3.9 Extending LANs: Repeaters and Bridges

Often the need arises to extend a DDC LAN from one mechanical room or building to the next. If distance constraints permit, the easiest and least costly technique is usually to extend the LAN through the use of a *repeater*. A repeater is a bi-directional device that connects two LAN segments, or subnetworks, at the physical layer by simply retransmitting whatever it hears on either side out the other. In the process, the signals are usually regenerated and amplified. The interconnected subnetworks function exactly like a single, larger network in every respect. No duplication of addresses is allowed and the same medium access control method applies to all stations uniformly: they all hear the same master if it is a master/slave network; each other if it is a CSMA/CD network; or share a common token if it is a token-passing network. The presence of the repeater is "transparent" to all participating devices, i.e., there is no way to even know it is there.

While repeaters interconnect subnetworks at the physical layer, bridges are devices that operate at the data link layer. As with networks extended using repeaters, the devices in the resulting network must all have unique data link or MAC addresses. A major difference is that each of the interconnected subnets continues to function autonomously in terms of medium access control. Another difference is that bridges may perform traffic filtering and may interconnect subnetworks that use the same medium access control strategy but different media.

In practice, several types of bridges are available. The simplest, and least efficient, pass all traffic bi-directionally, just like a repeater. Other bridges perform filtering based on the destination MAC address. The least sophisticated filtering bridges must be configured manually using some type of network management software on a connected workstation. The configuration process involves setting up a table in the bridge indicating which addresses belong to which subnet. The bridge can then refer to the table to determine which messages it must relay and which messages it can safely ignore since the source and destination devices are on the same subnetwork.

More sophisticated, and expensive, bridges "learn" by listening to all messages that pass by and then build their own address tables. If the budget permits, self-configuring bridges are obviously the way to go.

6.4 THE NETWORK LAYER

A principal advantage of repeaters and bridges is that no additional protocol control information needs to be added to a message for it to be passed along to its destination. Sometimes, however, matters are more complex. Suppose, for example, we need to set up communications between devices on subnetworks that use different medium access control strategies with different length addresses or on subnetworks of the same medium access control type but with redundant addresses. Examples of this would be setting up communications between machines on Ethernet and ARCNET segments, respectively, or between machines on two ARCNET segments, each of which has the same address on its own segment. This function has been assigned to the layer in the protocol stack called the network layer. Devices that perform this type of relay function are usually called routers.

6.4.1 Extending LANs: Routers

Routers perform their relaying function by implementing some form of network layer protocol which inevitably adds more overhead in the form of protocol control information to the original message. In the simplest case, this may amount to nothing more than a field containing a *network address*. The network address works like an area code in the telephone system: it provides a way to distinguish between two parties with the same phone numbers in their respective areas (same physical device address) or with phone numbers of differing numbers of digits (different types of media access control).

Routers got their name because one of their main tasks is to figure out how to route messages! Two mechanisms are in use. The first, and simplest from the router's perspective, is called *source routing*. In this case, the originator of the message provides the route that is to be followed from the source to the destination in the form of a list of nodes through which the message must pass. All the router has to do is look down the list, see which device is next in line, and send on the message. This is the technique used by 802.5 token-ring routers.

The alternative to source routing is the use of look-up tables in each router that indicate, for each network address, the possibilities for further relay. In a sense, this is where the complexity of network layer protocols enters in. If there is only one path from source to destination, no decisions need to be made. If alternative paths exist, however, then some type of decision-making algorithm must be used to figure out the best route. Some protocols use the concept of "cost." When the routing tables are set up, a cost is assigned to each possible route based on factors such as the number of hops, bit rate, expected quality of the physical connection, and so on. By adjusting the cost value, which is usually not in dollars but in some arbitrary scale from 1 to 100, for example, it is possible to bias the routers to preferentially select the most desirable, efficient route.

Other possible network layer functions include segmentation and reassembly of messages, flow control, and expedited data transfer. The latter can be important in DDC or other process control applications. Suppose a router is operating between two subnetworks of different speeds. The messages arriving from the faster subnet will have to be stored for awhile before the slower subnet will be able to receive them. If an important message arrives from the faster side, e.g., a critical alarm message, the network layer protocol may need to have the capability of deviating from a first-in, first-out policy in its relay function. One way of doing this is by assigning a network layer priority to each message which the router can use to juggle the order of messages in temporary storage, moving lower priority messages down in the queue. Another way is to use a "special delivery" message format such that any time such a message arrives it is simply moved ahead of all other messages, except perhaps other special deliveries.

6.5 THE TRANSPORT, SESSION, AND PRESENTATION LAYERS

Few of the functions assigned to the three layers between the network and the application layers are needed in the types of local area networks used in DDC and other forms of process control. As a result, most DDC networks are built on a *collapsed architecture*, usually consisting of the physical, data link, network, and application layers. Nonetheless, in the interest of completeness, let's take just a moment to look at a few of the most important functions of these intermediate layers.

The transport layer is usually described as the first layer in the protocol stack that has "end-to-end" significance. The reason for this is that the OSI model was originally framed with wide area networks in mind and tends to presume that end systems are not on the same local network. Thus, the lowest three layers handle communication on the local segment and the routing of messages between multiple subnetworks while the upper layers are more oriented toward the requirements of the cooperating applications in the source and destination machines.

Probably the most important task of transport layer protocols is to handle the segmentation and reassembly of messages and assure their proper sequencing. This can be critical in wide area networks where routers may send successive messages over different paths, possibly operating at different speeds. In such a case, there exists the possibility that the messages, or message segments, may arrive out of their original order or that some segment might not make it through at all. Detecting such events, the transport protocol can take the necessary recovery actions such as requesting retransmission of the missing segment. Since most DDC messages are short and self-contained, this functionality is rarely needed.

Session layer protocols are most useful for communications which are characterized by lengthy data exchanges which cannot be readily repeated from the beginning if the communication link fails. One example would be file transfers. One of the main services provided by the session layer is to insert synchronization points in the data stream so that the transaction may be restarted at a known, specific place in the event of a failure. The session layer can also be used to establish a particular form of dialog, either half-duplex or full-duplex, i.e., one-way alternate or two-way simultaneous. Note that these terms are also used in a totally different context to describe interactions at the physical layer. Another function of the transport layer is to respond to changes in the "quality of service" of the underlying network. Again, this is not really an issue with LANs which tend to work nearly flawlessly or not at all!

One of the goals of the OSI model is to provide a framework for developing protocols leading to interoperability. A main concern of the developers was the use of different character codes to represent the same information. Two of the main contenders were the ASCII and EBCDIC codes discussed earlier. To address this problem, the OSI model allows for the possibility of data transformations from one form to another. At the application layer, for example, a data element consisting of text may be represented abstractly as a *character string*. It is the role of the presentation layer to negotiate a representation of the text string with its peer presentation layer in the cooperating machine and then perform the necessary encoding, including possibly data compression and data encryption. If, however, the representation of the data is agreed to in advance and no data transformations are required, an explicit presentation layer protocol is not needed. This is the normal situation for DDC communications.

6.6 THE APPLICATION LAYER

Finally we come to the highest, and in many respects the most important, layer in the protocol stack. It is the function of the application layer to provide the services required by cooperating applications in different computers. Since each such application—whether word processing, computer-aided drafting, airline reservations, automatic teller machines, *or* DDC—has highly application-specific communication requirements, each industry has had to wrestle with its own individual standardization effort. In the case of the building automation and control industry, this has led to the development, through ASHRAE, of BACnet, a comprehensive protocol specification that will be described in the next chapter.

The interaction between a DDC application process and the communications services of the application layer is carried out by means of an *application program interface,* or API. In practice, the API usually consists of a subroutine call, a function call, or a procedure call, depending on the language in which the DDC operating system is written. Whatever the specific mechanism, parameters are provided to the communication software which the application services can use to build messages appropriate to the required communication.

If you have worked with any DDC system, you can probably make a reasonable list of the services that need to be provided. First are undoubtedly services to read and write the values of binary and analog points. "Writing" to a hardware point is another way of saying commanding the point to a given state or value. There may also be a need for a service to get at information directly related to a point such as its perceived reliability, error status, time of last value change, and so on.

Another important class of services involves communicating data about alarm conditions and other events. The latter include binary changes of state or analog changes of value, usually in excess of some predefined increment, which may need to be known by another control device somewhere on the network. A human operator may want to be able to invoke messages (through the OMI interface) that request a summary of the status of the alarm conditions defined in a remote device or, perhaps, "enroll" the operator to be notified about certain other events of interest when they occur.

A very common requirement is to be able to transfer programs and database information from one machine to another. Often this function can be accomplished by means of file transfer services. Sometimes, however, additional actions are required on the part of the DDC such as refraining from changing the state of output points during the time a control program is being updated. Such constraints are often part of "upload" or "download" services which not only accomplish data transfers but may do so in a way which averts any disruptive side effects.

Other services may be defined to handle time-related functions such as event scheduling and time synchronization, and various hardware-related device management functions such as the performance of self-test diagnostics or the reloading of the device's software from ROM.

Some DDC devices support virtual terminal communication services. The virtual terminal concept allows information to be exchanged across the network that can be used to make it appear that an operator workstation is directly connected to a DDC device through a dedicated physical connection at the DDC site. This may allow an operator to perform certain types of configuration and diagnostic operations that would otherwise require a visit to the mechanical room.

From the foregoing discussion two points should be clear: (1) it is the application layer services that carry out the bulk of the communication processing relating to DDC; and (2) given the diversity of different vendors' systems, it is not surprising that the road to a standardized way of performing these services has been a bumpy one!

6.7 THE FUTURE OF NETWORKING

In many respects, the power of direct digital control systems lies in the networking technology which allows information on the conditions in the farthest-flung corners of a building, or building complex, to be more or less instantly available. Recognizing that the bulk of this technology is *not* the product of the building automation and control industry but rather the computer industry at large, what does the future hold in store for DDC networks?

First, let's look at the state-of-the-art. Lower layer protocol standards are now abundant, well-proven, and increasingly affordable. The effect of this is that the means are now available to interconnect DDC devices using nearly any medium at any communication speed. You have but to look in the trade literature to see the increasing number of vendors offering Ethernet, ARCNET, or fiber optic network options.

Second, there is an increasing sentiment in favor of vendor-independent interoperability, not only in the computer world at large but specifically in the realm of DDC. This is manifested in both the increasing number of vendors who have chosen to publish their protocols, thus making them "open" to use by others, but also in the work of committees in the United States and abroad seeking to develop *standards* for DDC communication.

The fruition of these activities, exemplified by the BACnet effort described in the next chapter, will result in the capability to design and install networks of direct digital control components that will, in conjunction with the information display and interpretation capabilities under development in the operator workstation field, truly revolutionize the way buildings are managed and controlled.

REFERENCES

ANSI/ATA 878.1 (1992), *ARCNET Local Area Network Standard*, New York: ANSI.

EIA-232-D, *Interface Between Data Terminal Equipment and Data Communication Equipment Employing Serial Binary Data Interchange*, Washington: EIA, 1987.

EIA-485, *Standard for Electrical Characteristics of Generators and Receivers for Use in Balanced Digital Multipoint Systems*, Washington: EIA, 1983.

Hoss, R., *Fiber Optic Communications Design Handbook*, Englewood Cliffs, NJ: Prentice-Hall, 1990.

ISO 7498, *Information processing systems—Open Systems Interconnection—Basic Reference Model*, Geneva: ISO, 1984.

ISO 8802-2, *Information processing systems—Local area networks—Part 2: Logical Link Control*, Geneva: ISO, 1989.

ISO 8802-3, *Information Technology—Local and Metropolitan Area Networks—Part 3: Carrier Sense Multiple Access with Collision Detection (CSMA/CD) Access Method and Physical Layer Specifications*, Geneva: ISO, 1992.

ISO 8877, *Interface connector and contact assignments for ISDN*, Geneva: ISO, 1992.

Martin, J., *Telecommunications and the Computer*, Englewood Cliffs, NJ: Prentice-Hall, 1976.

Martin, J., *Computer Networks and Distributed Processing*, Englewood Cliffs, NJ: Prentice-Hall, 1981.

McNamara, J., *Technical Aspects of Data Communication*, Maynard, MA: Digital Press, 1988.

Stallings, W., *Handbook of Computer-Communications Standards*, Volumes 1, 2, and 3, New York: Macmillan, 1987.

Stallings, W., *Local Networks*, New York: Macmillan, 1990.

SOURCES FOR STANDARDS

ARCNET: ANSI (see ISO below) or ARCNET Trade Association, 3227 North Wilke Road, Suite 2706, Arlington Heights, IL 60004

EIA: Electronic Industries Association, 2001 Eye Street N.W., Washington, DC 20006.

ISO: American National Standards Institute, 11 West 42nd Street, 13th Floor, New York, NY 10036.

Chapter 7

ASHRAE's BACnet

SUMMARY

In June of 1987 a group of individuals from the building controls industry gathered for the first time in Nashville, Tennessee, to deliberate the creation of a communication protocol for the purpose of enabling computer-based building automation and control equipment to interoperate. The individuals came from manufacturers, universities, consulting engineering firms, and governmental agencies. The group was formed under the auspices of ASHRAE and was called Standards Project Committee 135P (SPC 135P). I had been previously asked to chair the committee.

Considerable controversy surrounded the initial meeting. Would the manufacturers really cooperate with what many viewed as a user-inspired effort? What would be the scope of the protocol? Would individual vendors fight to have their own protocols adopted as the standard? Would the chairman live to tell the tale?

Happily, the answers to these and similar questions have been most encouraging. In the fullness of time, participants in the work of the SPC have all come to realize the potential benefits of standards, particularly standards arrived at by the painstaking process of concensus-building. The results, of course, contain compromises. But many of the protocol's features represent the distilled wisdom and experience of some of the industry's most dedicated communications professionals.

One such concensus effort was to find a simple, easily remembered name to identify what for nearly three years had been referred to merely as "the ASHRAE protocol." This was not as easy as you might think. We obviously could not call the standard "ASHnet" and "BACtalk" was rejected as just a bit too sassy (although there was considerable support for it by virtue of its analogy to "AppleTalk" and "DECtalk"). In the end, we agreed on "BACnet," an acronym that stands for *B*uilding *A*utomation and *C*ontrol *Net*work. The name seems to properly reflect the hope of the committee that the protocol will ultimately facilitate the interoperability of not only DDC equipment but also, potentially, that of other common types of building automation equipment.

In this chapter I will try to both explain BACnet as it exists today and give you some insight as to the underlying design philosophy that has shaped it.

7.1 WHAT IS BACnet?

Having recognized the need to develop a standard protocol, the first issue that had to be addressed was "how." We decided that there were four key components to the development process.

The first was to solve the problem of how to represent the internal functioning of any vendor's equipment in a common, "network-visible" way, recognizing both the proprietary nature of each vendor's internal design and the diversity of functionality involved.

The second key component was to agree on a set of common commands or services that could be used between devices to get them to carry out the functions that we thought would be required to implement distributed monitoring and control. In other words, what kinds of messages should the protocol provide to facilitate communications about common building automation functions and what rules should govern their exchange?

The third piece of the puzzle was to agree on how to encode the messages defined above in a standard way. How should the messages be represented as binary zeros and ones on the communications media?

Finally, what LAN technologies should be used to actually get the BACnet messages from one device to another. While many observers thought that this last issue would have to be addressed first, the SPC disagreed. Our approach was: "let's figure out what the devices will say to each other, *then* worry about how to get the messages there, probably with LAN technology developed outside of ASHRAE." In terms of the OSI model, this was a "top-down" approach since our initial focus was on establishing the functioning of the application layer. As a result, discussions on the lower layers were deferred until the message structure, content, exchange rules, and encoding of the application layer had pretty well been decided. These discussions eventually led to the adoption of a collapsed OSI architecture consisting of application, network, data link, and physical layer functionality. Had we *not* taken this path, we would probably still be arguing over wire types and baud rates!

Let's look now in more detail at each of these four fundamental aspects of the BACnet specification.

7.2 MODELING DDCs AS NETWORK-VISIBLE OBJECTS

The first challenge for the SPC was to come up with a way of representing the internal functioning of disparate devices in such a way that it could be the subject of a computer-to-computer dialog over a network. To accomplish this, we looked initially at DDC devices. We reasoned that most such devices perform very similar tasks even though their internal software data structures are likely to be significantly different.

The trick would therefore be to "map" the internal data structures to common, well-defined abstract data structures that we decided to call *objects.*

The idea behind an object is that it is a collection of data elements, all of which relate to a particular function. The individual data elements are called the *properties* of the object. While the internal design and configuration of building automation devices will most likely continue to be proprietary and therefore different for each vendor, the properties of objects provide a means of identifying and accessing information without requiring any knowledge of the details of a device's internal design. The communication software in the device can interpret requests for information about these abstract objects in such a way as to obtain the information from the real data structures inside the device. These objects collectively provide a *network-visible* representation of the building automation device. See Fig. 7.1.

BACnet presently defines 18 different object types as summarized in Table 7.1. Any given building automation device may have zero, one, or many objects of each object type. The key to accessing a particular object is a property called the *object identifier* which uniquely identifies each object within a specific device. The object identifier can be thought of as the "name" of the object, although it is not text but rather a 24-bit octet string containing a 1-bit field indicating whether the object type is defined in the standard or is proprietary, a 9-bit object-type field, and a 14-bit instance number. The object types described below represent most of the kinds of information found in modern DDC devices. As we will discuss later, additional object types may be easily added to facilitate communications about non-DDC devices that deal with other building automation functions such as fire, security, and lighting control. For each standard object type, BACnet defines a set of properties, the purpose and data type of which are described in detail in the standard. Table 7.2 shows, for example, the properties that are defined for Analog Input objects.

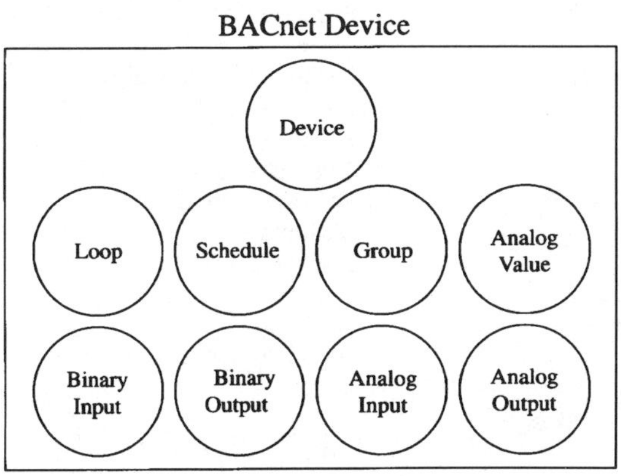

FIGURE 7.1. A BACnet Device may be thought of as a collection of objects, each of which has a specified set of properties. Which objects are needed to represent the underlying physical device depends on its functionality. The representation may be extended by defining additional objects as new functions are implemented or as new technologies appear.

TABLE 7.1 BACnet Standard Object Types

Analog Input	Event Enrollment
Analog Output	File
Analog Value	Group
Binary Input	Loop
Binary Output	Multistate Input
Binary Value	Multistate Output
Calendar	Program
Command	Recipient Table
Device	Schedule

Here is a brief description of the standard BACnet object types:

Analog, Binary Input — Represents the characteristics of physical analog and binary sensor inputs.

Analog, Binary Output — Represents the characteristics of physical analog and binary actuator outputs.

Analog, Binary Value — Represents the characteristics of analog and binary data values that are defined in software. Their values may be changed by BACnet services or by processes internal to the device in which they reside.

Calendar — Represents a list of dates which have special meaning when scheduling the operation of mechanical equipment. A list of holidays would be an example.

TABLE 7.2 Properties of the Analog Input Object Type

Property Identifier	Property Data Type
Object_Identifier	BACnetObjectIdentifier
Object_Typer	BACnetObjectType = ANALOG_INPUT
Present_Value	Real
Description	CharacterString
Device_Type	CharacterString
Status_Flags	BitString
State	BACnetReliability
Out_of_Service	Boolean
Update_Interval	Unsigned
Units	BACnetEngineeringUnits
Min_Pres_Value	Real
Max_Press_Value	Real
Resolution	Real

Data types beginning with "BACnet" are defined in the standard and are usually either enumerated values or allow for several possible forms. For example, a property of type BACnetEventState is an enumeration that takes on values or 0 to 5 corresponding to states of Normal, Alarm, Low-Limit, High-Limit, Low-Alarm, and High-Alarm.

Command	Allows a set of predefined commands to "fan out" in response to a value written to its present value property. For example, when the present value property is set to a value corresponding to "occupied," the lights could be turned on, the heating system enabled, and so on.
Device	Contains general information about a particular BACnet device, such as vendor name, model name, location, and object types supported.
Recipient Table	Provides a shorthand way to reference a list of devices. It may be used, for example, to reference several operator workstations or printers that must be notified when an alarm or other event occurs. It can also contain "raw" addresses to indicate that a broadcast message is to be sent.
Event Enrollment	Provides a way to define alarms or other types of events and to indicate which devices should be notified when they occur.
File	Provides information for the access and manipulation of files which are modeled as octet strings or records of arbitrary content.
Group	Provides a shorthand way to read a number of properties with one read request. It might be used, for example, to update several fields in an operator graphic display with a single communication.
Loop	Can be used to represent any feedback control loop which is some combination of proportional, integral, or derivative control.
Multistate Input	Represents the characteristics of a physical or logical entity which can take on several different states. For example, a fan system might have states OFF, SLOW, FAST.
Multistate Output	Represents the desired state of one or more physical outputs and/or processes within the BACnet device in which the object resides.
Program	Represents the network-visible characteristics of an application program or process without specifying the logic it contains. By manipulating properties of objects of this type, programs may, for example, be started or stopped.
Schedule	Provides a means to cause the writing of a specific value to a specific property of any given object based upon time-related criteria.

Having looked through the list of standard object types, you may be wondering how objects are set up in an actual device. In the arcane jargon of standards committees, this is a "local matter," i.e., not specified in the standard and left to the discretion of each implementer. In practice, the most common way will probably be through the use of software that is executed at the time a particular device is configured or programmed. It may also be possible to create certain objects using the BACnet CreateObject service or interactively using the virtual terminal services described in the next section.

7.3 BACnet SERVICES

Objects provide an abstract representation of the network-visible portion of a building automation device. The *application services* provide "commands" for accessing and manipulating this information as well as providing some additional functions. BACnet currently defines 29 application services which are grouped into five categories. See Table 7.3.

In addition to the categories shown in the table, services are further classified as "confirmed" or "unconfirmed." A confirmed service results in a return message from the device that receives the service request called an acknowledgment or ACK. The ACK may be either simple or complex. A simple ACK contains no information other than that the service was successfully carried out. A complex ACK contains additional information which varies from service to service. An unconfirmed service is not acknowledged but can be useful in certain circumstances, for instance, to disseminate a periodic time synchronization message to all devices on the network.

In order to exchange these messages, communication rules in the form of an application layer protocol also had to be developed. What emerged was a commun-

TABLE 7.3 BACnet Application Services

Alarm and Event Services	Remote Device Management Services
AcknowledgeAlarm	ConfirmedPrivateTransfer
ConfirmedEventNotification	UnconfirmedPrivateTransfer
GetAlarmSummary	ReinitializeDevice
GetEnrollmentSummary	ConfirmedTextMessage
UnconfirmedEventNotification	UnconfirmedTextMessage
	TimeSynchronization
File Access Services	Who-Is
AtomicReadFIle	I-Am
AtomicWriteFile	Who-Has
	I-Have
Object Access Services	
AddListElement	Virtual Terminal Services
CreateObject	VT-Open
DeleteObject	VT-Close
ReadProperty	VT-Data
ReadPropertyConditional	
ReadPropertyMultiple	
RemoveListElement	
WriteProperty	
WritePropertyMultiple	

ication model based on the idea of a client-server relationship as shown in Fig. 7.2. The client sends a service request to a server which attempts to carry out the service and provide some kind of response.

In BACnet's case, the implementation of this model required the definition of eight distinct message types or *protocol data units* (PDUs) each of which consists of *protocol control information* (PCI) and, possibly, service-related data. Because this part of BACnet concerns the application layer of the OSI model, these PDUs are called application layer PDUs or APDUs. (Other OSI layers produce their own PDUs, e.g., network layer PDUs are called *N*PDUs, data link layer PDUs are called *L*PDUs, etc.) The eight BACnet APDU types are:

BACnet-Confirmed-Request-PDU
BACnet-Unconfirmed-Request-PDU
BACnet-SimpleACK-PDU
BACnet-ComplexACK-PDU
BACnet-SegmentACK-PDU
BACnet-Error-PDU
BACnet-Reject-PDU
BACnet-Abort-PDU

The use of the first four was hinted at above. The SegmentACK PDU is used by BACnet's segmentation/reassembly mechanism to acknowledge the receipt of a segment of a message that had to be broken up due to limitations in the maximum message size imposed by the underlying LAN. Such lengthy messages might occur if an operator asked for details about all the objects in a particular device or a summary of all the current alarms (let's hope the last example is strictly fictitious!). The error and reject PDUs are used to convey information about problems in carrying out or interpreting a previously transmitted PDU. The abort PDU is used to terminate

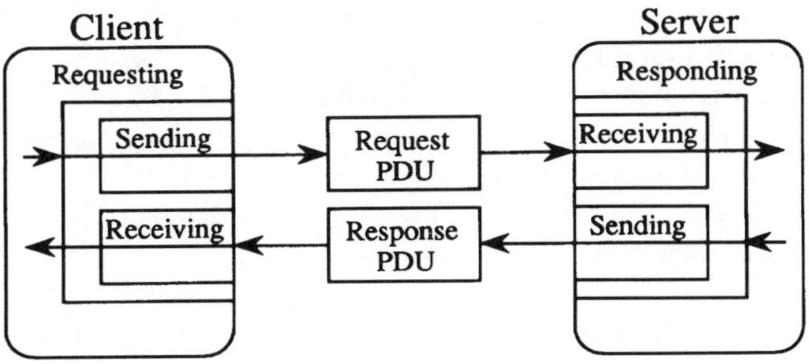

FIGURE 7.2. BACnet makes use of the Client-Server model of communication. The Client device request a service from a Server device by sending a Request PDU. The Server responds, if a response is called for by the nature of the requested service, with a Response PDU.

a transaction due to the emergence of resource limitations or other unexpected problems.

The PCI contained in these APDUs is used by a requester to match up a response with a previous confirmed service request; to control the segmentation/reassembly mechanism; and to indicate which of the various BACnet services, described next, is being requested or responded to.

7.3.1 Object Access Services

Of the five service categories, the object access services will probably get the most use, with the ReadProperty service likely to be invoked more frequently than any other. (In the standard, service and data type names are "scrunched" together because our syntax rules don't allow for spaces in such names. Now that you've been warned, I'll follow that convention here!)

Let's look at how a ReadProperty transaction works. Let's assume that an application program needs the value of a certain analog sensor input. The application program could be, for instance, a control program or a program used by an operator to produce a graphic display. The program, whatever its function, contains logic to call the communications software that in turns generates BACnet messages and sends them out onto the LAN. The form this *call,* or *application program interface,* takes depends on the software environment in which the program is running. It could be a subroutine call in a FORTRAN environment, a procedure call in a Pascal environment, or a function call if the application program is coded in C. In any case, the call must pass certain parameters to the communications software. Among these must be: the name or address of the destination device; the name of the service that is being requested; whatever parameters are required for BACnet to carry out the service; and, perhaps, if the message is urgent, a priority value.

Refer to Table 7.4. This table, a somewhat streamlined version of the tables contained in the BACnet standard for each service, indicates what parameters are required both to request and to respond to the ReadProperty service. The letters in the Request and Response columns mean:

M = This is a mandatory parameter and must be supplied.

U = This is a user-selectable parameter which may, or may not, be present depending on how the service is being used.

C = This is a conditional parameter whose presence depends on the value of a previous parameter.

S = This is a selection parameter. The parameter is selected from two or more choices at the same table indentation level.

In the case of the ReadProperty service we see that the Argument parameter is required and consists of up to three subparameters. The first is the Object Identifier of the object containing the property whose value we wish to learn. The second subparameter is the Property Identifier. Finally, there is the Property Array Index

TABLE 7.4 Parameters of the ReadProperty Service

Parameter Name	Request	Response
Argument	M	
Object Identifier	M	
Property Identifier	M	
Property Array Index	U	
Result (+)		S
Object Identifier		M
Object Type		M
Property Identifier		M
Property Array Index		C
Property Value		M
Result (–)		S
Error Type		M

subparameter. It is user-selectable because it is only needed if the property referred to by the Property Identifier happens to be an array.

Assuming the ReadProperty request message is properly encoded and conveyed across the network to its destination without incident, the destination BACnet device would attempt to carry out the service request in accordance with the service procedure set forth in the standard. If it is successfully able to access the requested object and property, it produces a Result(+) response. If not, a Result(–) is generated. The subparameters provided in the Result(+) response are the same as contained in the Argument with the addition of the Object Type and Property Value. The latter, of course, is what our application program is trying to find out! If the ReadProperty service fails, the responding device sends back a Result(–) response indicating the reason. The most common reason would be that the specified object was not actually defined in the responding device or that an invalid property had been requested.

After ReadProperty, the WriteProperty service can be expected to be the next most heavily used of the Object Access services. It requires the same subparameters as those of ReadProperty with two additions: the Object Type of the target object must be specified and, if the specified Property Identifier refers to a commandable property, a priority value must be supplied. Knowing the object type will be helpful to vendors who want to build gateways to existing DDC systems. The priority concept is worth describing in more detail.

A major concern in the design of distributed control systems is coordinating the modification of states or values between the various control processes that may wish to make such changes. In the case of typical HVAC systems, there may be different control modes for emergencies such as the presence of fire or smoke, startup, temperature override, demand limiting, optimum start/stop, time-of-day scheduling, and manual control. In addition, there needs to be a mechanism for resolving the competition between processes local to a device and those that reside in remote devices and want to make state or value changes using BACnet services. The commandable property mechanism solves this problem.

By definition, a commandable property is any property which is writable using BACnet services and is simultaneously subject to the BACnet command prioritization mechanism. This mechanism requires the establishment of two additional properties for each commandable property. They are the Priority_Array and the Relinquish_Default (default value) properties. In the current version of BACnet only the Present_Value properties of the Analog Output and Value, Binary Output and Value, and Multistate Output objects are commandable.

Prioritization works like this. The priority array consists of 16 elements, or "slots," each of which is associated with a priority value from 1 to 16, 1 being the highest. Each priority value is also uniquely associated with an application such as, for example, emergency control or temperature override. At any given time, each slot in the priority array contains either a legitimate value for the commandable property or a NULL value. The object management software in the BACnet device is responsible for periodically scanning the priority array and assigning the highest non-NULL value to the commandable property. If all the slots contain NULL, the default value is assigned to the commandable property.

To issue a command, a local process simply fills in the appropriate slot with the desired value. A process in a remote BACnet device issues a WriteProperty service request with the desired value and priority. Upon receipt of the service request, the local BACnet device fills in the appropriate slot with the newly received value, overwriting whatever was there. If the slot previously contained NULL and is a higher priority slot than the slot previously in control, the new value "takes over" the commandable property (subject, possibly, to a minimum on or off time constraint mechanism that the standard also defines). If the slot is below the slot currently in command, nothing happens until the higher priority application relinquishes control. A remote application relinquishes control of a slot by issuing a WriteProperty service request to the commandable property at the appropriate priority using a property value of NULL.

In addition to ReadProperty and WriteProperty, each of which accesses only one property at a time, there are also services to read and write more than a single property as the result of a single service request. These are called ReadPropertyMultiple and WritePropertyMultiple. One note about the multiple write service: in the early drafts of the standard it was going to be an "all or nothing" service meaning that the service request would have failed, in its entirety, with no properties modified, if any one of the specified properties had not been successfully accessed. The problem with this is that it would have placed a significant "rollback" burden on the service attempting to carry out the service. We could also have defined the service to try to write "as many as possible." In the end, we decided to define the service such that it writes "up to the first property that fails" and reports back to the requester information about which property has problems.

The AddListElement and RemoveListElement services allow the modification of properties that are complex in that they consist of a list of values. Common examples are the notification recipient lists of Event Enrollment objects or the list of object-property references in a Group object. These services were added because the simple WriteProperty service only allows properties to be modified in their entirety.

A subject of ongoing debate in the SPC has been where to draw the line between functions that can only be carried out by proprietary configuration techniques and those that can be performed using BACnet services. The creation of new objects was such a function. As mentioned above, objects are generally created by a configuration process unique to each implementation of BACnet and outside the scope of the standard. But some objects really do need to be created dynamically, on the fly. As a result, we decided to add the CreateObject service which can create whatever object types a vendor wishes to support. In order that a prospective purchaser can know what object types these are, the vendor must provide a list of the supported object types in each device's Protocol Implementation Conformance Statement (described later). As a conjugate to CreateObject, BACnet also provides a DeleteObject service to make possible the deletion of objects created dynamically.

The last object access service is ReadPropertyConditional. In some ways, it is the most complex service defined in the standard. The idea, however, is simple: to cause the service provider to return a list of properties from all objects possessing one or more properties meeting certain logical conditions. To accomplish this, each ReadPropertyConditional request contains object selection criteria and a list of properties that are to be returned from each selected object. The selection criteria consist of Boolean expressions of the form <property value-relation specifier-comparison value> along with a parameter that indicates whether the Boolean expressions should be AND'ed or OR'ed together.

As an example, ReadPropertyConditional could be used to return the Object_Identifier, Present_Value, and Reliability properties of all objects with <Object Type – equal to – Analog Input> AND <Present Value – greater than – 100>. Similarly, the service could return the object identifiers of all objects with <Reliability property – not equal to – NO_FAULT_DETECTED>. The potential uses for this service are nearly limitless.

7.3.2 Alarm and Event Services

Probably the second most important class of BACnet services are those that relate to communication about events and the special class of events known as *alarms. Events* are changes of value of any property of any object that meet predetermined criteria. The object whose properties are referred to is known as the reference object. The criteria used to ascertain that an event has occurred are determined by the event type. The list of those devices that are to be notified when a particular event occurs are contained in an Event Enrollment object for the specific event.

BACnet recognizes two categories of events: alarm events and nonalarm events. The primary distinction between the two categories is the intended recipient of the event notification. Alarm events are intended to be received and, optionally, acknowledged by a human operator. Nonalarm events are primarily intended to be of interest to peer monitoring or control programs.

Eight event types are specified in the standard because of their widespread occurrence in building automation and control systems:

1. CHANGE_OF_STATE_ALARM
2. CHANGE_OF_VALUE_ALARM
3. COMMAND_FAIL_ALARM
4. FLOATING_LIMIT_ALARM
5. OUT_OF_RANGE_ALARM
6. CHANGE_OF_STATE_EVENT
7. CHANGE_OF_VALUE_EVENT
8. OUT_OF_RANGE_EVENT.

Other event types may be defined on a vendor-specific basis. In addition, the occurrence of events resulting from the execution of nonstandard algorithms may be signaled by means of one or more of the eight standard mechanisms applied to the Present_Value property of binary or analog value objects. The practical effect of this last point is that *any* algorithm can be used to produce an alarm or event notification, not just the standard ones.

Events in BACnet are based on algorithms applied to the properties of specific objects called reference objects. Such a property is called a *referenced property* and is one of the properties of an Event Enrollment object. The parameters required to compute the current state of the event are contained in the parameter list property of the same Event Enrollment object. In addition, each Event Enrollment object contains a list of *notification rules* which indicate the state transitions, e.g., NORMAL_TO_ALARM or ALARM_TO_NORMAL, which should generate notifications to those devices which have "enrolled" as well as a parameter to indicate whether a human operator has acknowledged the occurrence of the event. As an example, consider the algorithm for an Out_Of_Range_Alarm event. According to the standard:

> An OUT_OF_RANGE_ALARM occurs if the Referenced Property leaves a range of values defined by the High_Limit and Low_Limit parameters. If the transition is to a value above the High_Limit, the NORMAL_TO_HIGH_ALARM transition is used to determinine if an event notification is to be generated. If the transition is to a value below the Low_Limit, the NORMAL_TO_LOW_ALARM transition is used.
>
> An OUT_OF_RANGE_ALARM clears when the Referenced Property attains a value greater than the (Low_Limit + Deadband) or a value less than the (High_Limit – Deadband). (Note that the limit values may be Boolean TRUE or FALSE as well as analog values.) The transition used to determine if an event notification is to be generated is LOW_ALARM_TO_NORMAL or HIGH_ALARM_TO_NORMAL depending on whether the return to normal is from below the Low_Limit or from above the High_Limit.

Once an event has occurred, one of two services is used to communicate the news:

the ConfirmedEventNotification or the UnconfirmedEventNotification. The main difference between the two services is that the former must always be directed to a specific recipient while the latter may be broadcast. If the event is an alarm, a human acknowledgement may be desired. This is conveyed by means of the Acknowledge-Alarm service and has the effect of setting the Boolean flag in the corresponding Event Enrollment object. Note that a ConfirmedEventNotification results in an acknowledgment from the receiving device that the notification has been received by the machine, not necessarily seen by a human.

Two additional services have been defined to help with the management of events and alarms. They are particularly useful for communicating with simple devices that may not support the ReadPropertyConditional service. The first is the GetAlarmSummary service. It causes the service provider to search all its Event Enrollment objects and return the object identifiers of those whose event type is one of the ALARM types listed above and which have a state property not equal to NORMAL. The acknowledged state, i.e., whether an acknowledgment from a human operator has been received, is also returned in the summary.

The second additional service is called GetEnrollmentSummary. You might think of it as sort of a ReadPropertyConditional service specifically tailored to work on Event Enrollment objects. The argument parameter contains up to five "filters" which are really conditions that an Event Enrollment object's properties must meet in order to be included in the summary. These include: for alarm events, whether the event has been acknowledged by a human; whether a particular device is included in the notification recipient list; and whether the event is in a particular state. Additional filters let you limit the summary to only those Event Enrollment objects of a certain type or with a certain event priority value.

7.3.3 File Access Services

An important DDC communication requirement is to be able to transmit and receive files. As used in BACnet, a *file* is a network-visible representation of a collection of octets of arbitrary length and meaning. This is an abstract concept only and does not imply the use of disk, tape, or other mass storage devices. Thus a file could be a contiguous area of memory as well as a collection of octets residing on disk. BACnet services may also be used to access vendor-defined files as well as specific files defined in the BACnet standard. The file access services can model files in two ways: as a continuous stream of octets or as a sequence of numbered records.

The effect of this definition is that the two BACnet file access services, AtomicReadFile and AtomicWriteFile, can be used to perform all kinds of data transfer including what is commonly called *upload* and *download.* Note that what these terms actually mean varies greatly from system to system. For example, a download may be nothing more than a conventional file transfer, usually from a workstation to a controller. But it may also involve the controller taking certain actions such as freezing all output values or clearing out certain memory buffers. In order to accommodate these various possibilities, the SPC came up with the idea of a File object. The File object is used to identify a particular file by name and to provide access to *header information* such as the file's total size, creation date, and type. Armed with this information, a controller can decide what actions to take, outside the scope of BACnet, when an effort is made to modify the contents of a specific file using BACnet services.

You may be wondering about the significance of the rather explosive prefix "atomic." Atomic means that during the execution of a given read or write operation, no other read or write operations are allowed for the same file. In the jargon of conventional file services, the file is opened, read or written, and closed all in one self-contained, atomic operation. Synchronization of these services with the internal operations of the BACnet device is a local matter and is not defined by this standard.

7.3.4 Virtual Terminal Services

At first hearing, the term *virtual terminal* definitely has a strange ring to it! Yet the BACnet virtual terminal services will almost certainly find widespread use. Again, the idea is simple. You may remember back in the earlier chapters when we talked about going out to a mechanical room and plugging in some kind of terminal device directly into a controller for the purpose of configuration or looking at operational data. What the virtual terminal services let you do is conduct the same kind of operations but *over the network*. In other words, the virtual terminal services encapsulate the character-oriented keystrokes and responses that would normally pass through a dedicated cable between a terminal device, such as a laptop PC, and a DDC and allow them to co-exist on the same medium with whatever network traffic may be taking place at the same time.

In order to accomplish the virtual terminal function, BACnet provides three services: VT-Open, VT-Data, and VT-Close. Also, the Device object of devices that support VT services has two additional properties called VT Classes Supported and Active VT Sessions, the meaning of which will be described below.

In order to establish a VT session, the originating application program issues a VT-Open request specifying the desired type of terminal emulation or VT-Class. Each device capable of responding to a VT-Open request must support at least the VT-Class known as Default Terminal. This class is specified in BACnet and consists of a subset of ASCII including the printable characters and five of the most common control characters: BEL, BS, HT, LF, and CR. BACnet specifies in detail how a display device should respond upon receiving each of these characters. If a device is capable of other terminal emulation modes, e.g., Digital Equipment's VT-100 or Tektronix's 4010, these are made network-visible by means of the VT Classes Supported property of the Device object.

If the VT-Open service is successful, a VT-Session is set up and a unique VT-Session identifier is assigned to it. This numeric identifier allows the participating applications to unambiguously determine the session to which any particular VT message pertains. The session ID and the network address of the remote device are stored in the Active VT Sessions property of the Device object. The reason for this is primarily for error recovery. If a communicating partner "disappears," a system manager can find out what sessions once existed and then try to determine why the session was not properly closed.

Once a session is established, VT-Data messages may be exchanged containing whatever information is desired. When the session is complete, the VT-Close service can be invoked to terminate the communication.

7.3.5 Remote Device Management Services

Every standards committee needs a place to put its miscellaneous, but hopefully useful, concepts. In the case of BACnet, our miscellaneous services are collected together in the category we have called, for want of a better name, *remote device management services*. Let's look at what they do.

The Confirmed- and UnconfirmedPrivateTransfer services represent one of several "escape hatches" that we decided to build into BACnet so that down the road no one could legitimately claim that the protocol "stifled innovation." It lets a user transfer a group of octets without any regard for the contents whatsoever, although, of course, the sender and receiver must know what the data mean if any useful communication is to occur. These services can be used to accomplish all manner of potentially interesting, albeit nonstandard things. For example, a vendor could choose to use the PrivateTransfer services to send messages using an existing vendor-specific protocol over the top of BACnet. These messages could be directed to a gateway between a BACnet LAN and the proprietary LAN thus allowing a common, vendor-supplied workstation to communicate with both LANs over BACnet media. The PrivateTransfer services could also be used to carry out certain types of vendor-specific configuration activities outside the capabilities provided by the BACnet virtual terminal services.

The Confirmed- and UnconfirmedTextMessage services, like the confirmed and unconfirmed event notification services, provide a way to use BACnet to convey text of arbitrary content from one machine to another with, or without, acknowledgment from the receiving machine. Apart from the text itself, each message also contains a class, which may be a numeric value or text string, and a priority of either NORMAL or URGENT. The exact meaning of the class and priority parameters is left open so that implementers can do different things. One possible use of these services would be to send messages to a printer or log file.

The ReinitializeDevice service provides a way to reset a remote device to a known state (though this state may only be "known" in a vendor-specific way). The two states supported by this service as those that would be entered by virtue of a "warmstart" or a "coldstart." As you might imagine, trying to define these concepts in a vendor-independent way was not easy, though everyone agreed that BACnet should provide a reinitialization capability. Upon receipt of a warmstart request, possibly including an optional password parameter, a device shall reboot and "start over, retaining all data and programs that would normally be retained during a brief power outage." The precise interpretation of coldstart is left up to each vendor.

The TimeSynchronization service is an unconfirmed service which can be used to synchronize the time and date throughout the network.

The last BACnet services we will discuss are the potentially very important Who-Is, I-Am and Who-Has, I-Have services. In the case of Who-Is and I-Am, these services provide a mechanism for dynamically discovering the linkage between the symbolic name of a BACnet device and its network address. The device name is defined as the object name assigned to the Device object in a given BACnet device. It is required to be unique in a BACnet internetwork. (We will define the precise meaning of "internetwork" in Section 7.5.1). A BACnet network address consists of a network identifier and the medium access control (LAN) address of a device.

Who-Is works like this. Suppose an application knows that it wants to read certain properties of certain objects in a device called "Wiley Hall, Controller 1." If the local BACnet software knows the correlation between this name and the network address, it can simply direct the appropriate ReadProperty service to that destination. If, however, the correlation is *not* known, BACnet can try to find out the network address by issuing a broadcast Who-Is request with a Who-Is Device Name parameter of "Wiley Hall, Controller 1." If the indicated device hears the broadcast, it responds with an I-Am broadcast containing an I-Am Device Name parameter equal to the previous Who-Is Device Name parameter. By inspecting the source address information in the I-Am message, the originator of the Who-Is request, and anyone else who is interested, can build up a table correlating the device name with its address.

A device is also allowed to issue an I-Am message spontaneously. This can be used to notify potential communication partners that a device is back on-line after a lapse or to indicate a new network address. The same message has another fortuitous side effect. It can be used by BACnet routers to update their routing tables, a topic we will discuss in Section 7.5.1.

The Who-Has and I-Have services work the same way as just described except that Who-Has is used to locate instances of particular objects. The objects can be identified by using either the object's human-readable object name or its 3-octet object identifier.

7.4 BACnet ENCODING CONCEPTS

Having defined a set of objects, services, and APDU types, the next step was to figure out how to represent them in terms of zeros and ones. This process is referred to as *encoding*. For BACnet there were two major considerations: (1) how to represent the structure of a BACnet message so as to make clear the meaning of each part? and (2) how to convert a given message into a binary format suitable for transmission?

One possibility would have been to represent BACnet APDUs using the data structure definition formats of a standard computer language such as Pascal or C. We chose instead to make use of an international standard developed specifically for the purpose of describing PDU structures: ISO 8824, *Abstract Syntax Notation One (ASN.1)*.

The great advantage of ASN.1, and its associated encoding rules, is that it uses labels called *tags* to explicitly identify each PDU parameter rather than depending on an implicit knowledge of the parameter's position in a string of octets. This means that PDU encoders and decoders can be written that are not dependent on a particular predefined, usually static, arrangement of a PDU's contents but instead use the tags to figure out where one piece of data ends and another begins. New messages can therefore be added to the protocol with a minimum of additional programming. While the use of tags can increase the overall number of octets needed to convey a particular data element, it can also reduce the number of octets needed in the case where the presence of some of the parameters is optional. The main disadvantage of ASN.1 is that its rules are not yet as familiar to many people as the rules governing Pascal and C.

To understand the philosophy behind the "abstract" part of ASN.1 it is necessary to go back to the time when the OSI model was starting to be developed in the early 1980s. At that time one of the biggest data communication problems was the use of different character sets, primarily EBCDIC on the part of IBM and ASCII on the part of everyone else to represent the English alphabet. At the application level, the need was simply to transfer a character string with a certain value, e.g., "Hi, my name is Mike!"

To insulate the application program from having to deal with the formatting of the character string for transmission it was agreed to add a layer to the protocol stack (the presentation layer) to decide which character set to use on the transmission medium and to perform the actual encoding. The decision on which encoding to use would be established by negotiation by peer presentation layer implementations, and the encoding rules would then be applied to the data element whose type was represented by the *abstract syntax* to form a nonabstract sequence of bits called the *transfer syntax.* This negotiation could also be used to obtain agreement on other manipulations that might be carried out such as data compression or encryption.

While this is still the idea behind the presentation layer today, many protocols, such as BACnet, are based on architectures with fewer than all seven OSI layers. The absence of presentation layer functionality just means that communicating entities have to know which encoding rules are to be used by other means, the most obvious of which is "prior agreement"!

The agreement reached in the BACnet specification began with an investigation of the companion standard to ISO 8824: ISO 8825, ASN.1 Basic Encoding Rules. In this standard, each PDU data element is explicitly represented by three components: (1) tag octets, (2) length octets, and (3) contents octets, and the encoding is expected to apply uniformly to all components of a PDU. While the application of these rules produces an unambiguous encoding, the encoding efficiency for certain types of data is low. For example, the encoding of a single Boolean TRUE/FALSE value results in the generation of three octets: one that contains a tag value indicating that a Boolean value follows; one that contains the length of the following Boolean value; and one that contains a bit whose value indicates the Boolean state itself.

The alternative to such *explicit* encoding is to use *implicit* encoding where the position, length, and significance of a data element is known in advance. Although implicit encoding is almost always more efficient, you lose the extensibility that tags provide.

The approach taken in BACnet is a compromise and makes use of the fact that we have a limited number of APDU types; that every APDU has similar protocol control information; and that most of the data elements requiring transmission are likely to be only a small number of octets. The fixed portion of the APDU containing the PCI is encoded implicitly. The variable portion of the APDU containing service-specific information is encoded explicitly using a more streamlined encoding for tag and length values which allows us to reduce a Boolean value to a single octet. The tag and length values for other data will also fit in a single octet if the tag value is less than or equal to 14 and the length of the data is less than 7 octets. The result of all this is to reduce overhead while preserving the possibility of easily adding new application services in the future.

BACnet defines two types of tags: application and context specific. Application tags indicate that what follows is one of the data types defined in the standard and a given tag value always indicates the same data type. Application tags are encoded in a single octet:

7 6 5 4	3	2 1 0
Tag Number	c	L/V/T

where B7-4 = Tag Number corresponding to a particular data type. B'1111' indicates the tag number is in the next full octet.

B3 = Class bit. 0 = Application Tag, 1 = Context Specific Tag.

B2-0 = Length, Value, or Type of Encoding. Length of primitive data in octets (up to 4; 5 indicates length in subsequent octets), Value if data type is Boolean or Null, Type if following data has imbedded tags (6 indicates start, 7 indicates stop of constructed data).

The Tag Number field represents the BACnet application data types as follows:

Tag Number

0 = Null (L/V/T field = 0)
1 = Boolean (L/V/T = 0 = FALSE, L/V/T = 1 = TRUE)
2 = Unsigned integer
3 = Signed integer (2's complement notation)
4 = Real (ANSI/IEEE-754 floating point)
5 = Octet string
6 = Character string
7 = Bit string
8 = Enumerated
9 = Date
10 = Time

11 = Object identifier

12, 13, 14 = reserved for ASHRAE

15 = Type is found in the next octet; also reserved

Consider a simple 1-octet unsigned integer with a value of decimal 20 or hexadecimal 14. The encoding of the tag and data would be 2 octets long: X′2114′. The tag, X′21′ = B′00100001′. If you divide the fields according to the structure diagram above, you will find that the data type field contains B′0010′ = D′2′ = unsigned integer and the L/V/T field contains a 1 indicating that one octet follows.

Context-specific tags, i.e., tags with the class bit field set to 1, are used to identify data structures whose contents can only be deciphered if you know the details of the PDU structure, i.e., the "context" in which the tag is being used. A problem quickly arose when we considered implementation. Suppose you can't easily find out in advance how long a particular data element is? This could easily occur if a request was received to return the value of some parameter which happens to be a list of something. To address this problem, we use context-specific tags with the L/V/T field containing either a 6 or 7. The tag containing the 6 marks the start of the data while the tag with L/V/T field containing a 7 marks the end. You can think of these tags as bookends, between which a data element is sandwiched! This form of tag is also used when the following data is of known length but is "constructed," i.e., contains additional tags.

BACnet also defines a mechanism for extending both the tag number and length fields which is invoked if the tag number field is all 1's or the L/V/T field contains a 5.

As an example, let's consider a simple ReadProperty request. The ASN.1 is:

```
ReadProperty-Request ::= SEQUENCE {
    objectIdentifier        [0] BACnetObjectIdentifier,
    propertyIdentifier      [1] BACnetPropertyIdentifier,
    propertyArrayIndex      [2] Unsigned OPTIONAL
    }
```

This ASN.1 "production" illustrates several key points. First, it represents the definition of the ReadProperty-Request data type in terms of one of ASN.1's built-in data types called a SEQUENCE. A SEQUENCE is an ordered list of a specific number of elements. (There is also a SEQUENCE OF data type which is a list of *unspecified* length of a *single* type of data element). In this case, the SEQUENCE consists of three elements, one of which is OPTIONAL and may be omitted depending on the circumstances. Since all BACnet request and response productions are SEQUENCEs, we don't explicitly encode this fact, thus saving some overhead. Second, the numbers in square brackets are the context-specific tags described earlier.

The propertyArrayIndex parameter brings up a third point. From the ASN.1 we see that it is just an unsigned integer. If the array index is needed, it would be nice if we didn't have to have two tags, context tag [3] and a primitive tag indicating the presence of an unsigned integer, since we already know that the propertyArrayIndex is an unsigned integer *from the context*. The way around this is provided by the use of the ASN.1 IMPLICIT and EXPLICIT keywords. IMPLICIT means that the next tag, following a context-specific tag, may be omitted if the meaning is clear. Thus we could write, for example,

propertyArrayIndex [2] IMPLICIT Unsigned OPTIONAL

Because the use of IMPLICIT tags is so useful, ASN.1 lets you declare it as a default to be used without having to actually write IMPLICIT after each context-specific tag. This is what we have done in BACnet. Occasionally the default is overridden. This occurs if the data type of the following element is described by the ASN.1 CHOICE or ANY. This makes sense since otherwise the receiver of the message would have no way of knowing which data type was being conveyed.

Let's continue with our example. Before the APDU can be encoded, we need to know the ASN.1 for the data types represented by context tags [0] and [1]. The ASN.1 for BACnetObjectIdentifier is:

BACnetObjectIdentifier ::= OCTET STRING

where the contents of the 3-octet string are as described in Section 7.2.

Finally, the ASN.1 for a BACnetPropertyIdentifier is:

BACnetPropertyIdentifier ::= ENUMERATED {
acknowledged (0),
action (1),
action-text (2),
active-text (3),
•
polarity (65),
present-value (66),
priority-array (67),
•
vendor-name (104),
vt-classes-supported (105),
weekly-schedule (106),
window-size (107),
—Enumerated values 0-255 are reserved for definition by ASHRAE.
—Enumerated values 256 and above may be used by others subject

—to the procedures and constraints described in Clause 22.

}

The BACnetPropertyIdentifier is an ENUMERATED type where each property is represented by a number. In the interest of space, only a portion of the enumeration is shown above.

Since ReadProperty is a confirmed service, we will need to build a BACnet-Confirmed-Request-PDU. The format of a BACnet-Confirmed-Request-PDU is:

<table>
<tr><td>7</td><td>6</td><td>5</td><td>4</td><td>3</td><td>2</td><td>1</td><td>0</td></tr>
<tr><td colspan="4">PDU Type</td><td>SEG</td><td>MOR</td><td>SA</td><td>0</td></tr>
<tr><td>0</td><td>0</td><td>0</td><td>0</td><td colspan="4">Max Resp</td></tr>
<tr><td colspan="8">Invoke ID</td></tr>
<tr><td colspan="8">Sequence Number</td></tr>
<tr><td colspan="8">Proposed Window Size</td></tr>
<tr><td colspan="8">Service Choice</td></tr>
<tr><td colspan="8">Service Request
•
•
•</td></tr>
</table>

The PDU fields have the following values:

PDU Type = 0 (BACnet-Confirmed-Service-Request-PDU)
SEG = 0 (Unsegmented Request)
      = 1 (Segmented Request)
MOR = 0 (No More Segments Follow)
      = 1 (More Segments Follow)
SA = 0 (Segmented Response not accepted)
    =1 (Segmented Response accepted)
Max Resp = (0..7) (Maximum APDU size per table)
Invoke ID = (0..255)
Sequence Number = (0..255) Only present if SEG = 1
Proposed Window Size = (1..255) Only present if SEG = 1

Service Choice = (0..255)

Service Request = Variable Encoding.

Let's assume that our ReadProperty-Request will fit into one APDU. This means that the SEG and MOR bits will be 0 and that the Sequence Number and Proposed Window Size octets will be omitted entirely. The Invoke ID is a number assigned by the protocol to keep track of the response to the request. The service choice is derived from an enumeration similar to the one above. For ReadProperty-Request the value happens to be 10.

Let's assume for our example that we wish to read the Present_Value property of a Binary Input object with an instance number of of 3. For a standard binary input object, the 3-octet object identifier would be X'00C003'.

Taking everything into account the encoded message would be:

X'02'	PDU Type=0 (BACnet-Confirmed-Request-PDU), SEG=0, MOR=0, SA = 1
X'02'	Max Resp up to 256 octets (from table in standard)
X'11'	Invoke ID=17 (or some other value!)
X'0A'	Service Choice=10 (ReadProperty-Request)
X'0B'	Context Tag 0 (ObjectIdentifier, L=3)
X'00C003	Object Identifier
X'19'	Context Tag 1 (PropertyIdentifier, L=1)
X'42'	Property Identifier=66 (Present Value property)

The response to a BACnet-Confirmed-Service-Request is either a simple acknowlegment that the service was executed or a complex acknowledgment containing additional data. In the case of our example, we would expect to receive back a ReadProperty-ACK containing the requested property values conveyed in a BACnet-ComplexACK-PDU.

Armed with the information presented in this section, you should be able to get through the sections of BACnet that deal with encoding without too much trouble!

7.5 THE LOWER LAYERS

The lower OSI layers specified in BACnet are the network, data link, and physical layers. The network layer provides the means by which devices on different BACnet LANs can communicate. The data link and physical layers can be described in terms of the LANs themselves, two of which are the ARCNET and Ethernet LANs described in the previous chapter. In addition, BACnet provides a low-speed, low-cost option for "backward compatibility" with certain common types of existing equipment, much of which uses EIA-485 signaling. BACnet's EIA-485 LAN uses both master-slave and token-passing and is called MS/TP for short. There is also a mechanism for access to

BACnet networks over standard dial-up telephone connections. This section will give you a brief overview of these parts of the BACnet standard.

7.5.1 The Network Layer

To understand how the network layer functions, a few definitions are needed. These definitions are illustrated graphically in Fig. 7.3.

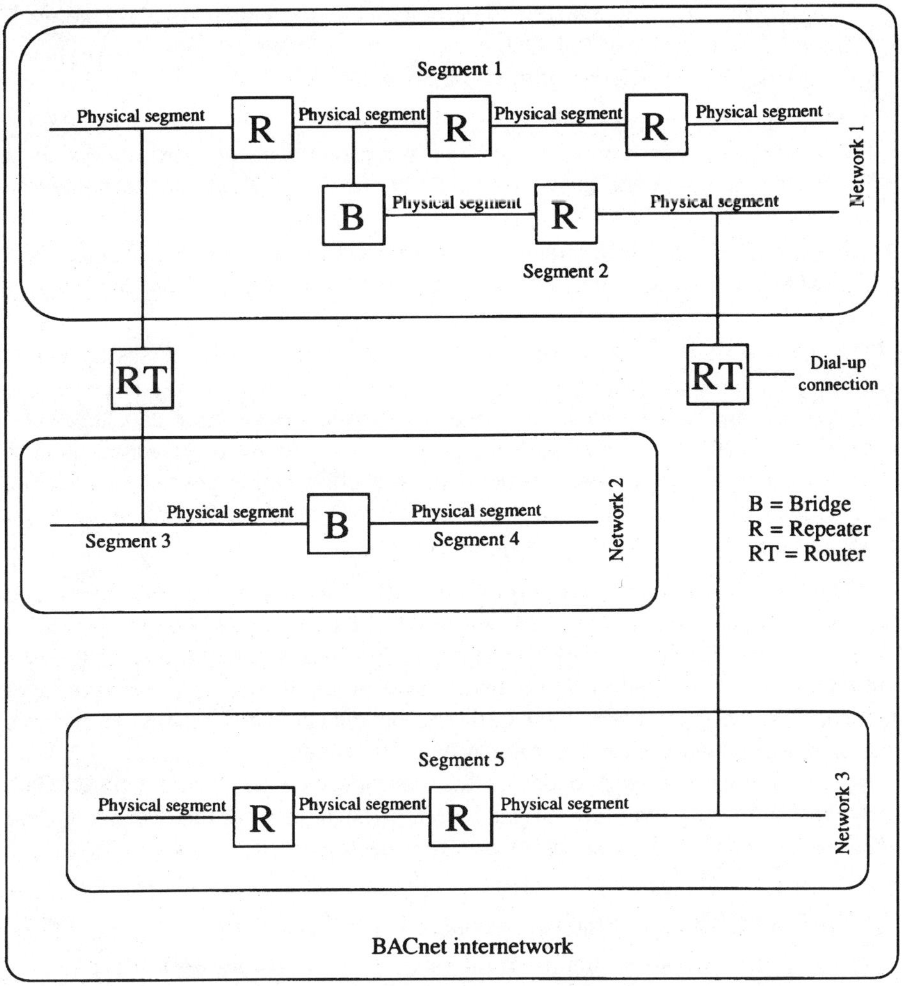

FIGURE 7.3. A BACnet Internetwork consists of two or more BACnet Networks, each consisting of one or more Segments, consisting in turn of one or more Physical Segments. The function of Repeaters, Bridges, and Routers is described in the text.

Physical Segment: A single contiguous medium to which BACnet nodes are attached.

Repeater: A device which connects two or more physical segments at the physical layer.

Segment: A segment consists of one or more physical segments interconnected by repeaters.

Bridge: A device which connects two or more segments at at the physical and data link layers. This device may also perform message filtering based upon MAC layer addresses.

Network: A set of one or more segments interconnected by bridges which form a single MAC address domain. Each network has a unique BACnet network number from 1 to 65534. Network number 65535 is used to address *all* networks for the purpose of a global broadcast.

Router: A device which connects two or more networks at the network layer.

Internetwork: A set of two or more networks interconnected by routers. In a BACnet internetwork, there exists exactly one message path between any two nodes.

BACnet Address: A pair of numbers consisting of a network number (2 octets) and a MAC address (6 octets for Ethernet, 1 octet for ARCNET and MS/TP).

The purpose of the BACnet network layer is to a provide the means by which messages can be relayed from one BACnet network to another, regardless of the BACnet data link technology in use on that network. Whereas the data link layer provides the capability to address messages to a single device or broadcast them to all devices on the *local* network, the network layer allows messages to be directed to a single device on a *remote* network, broadcast to all devices on a *remote* network, or broadcast globally to all devices on *all* networks.

Devices which interconnect two disparate BACnet LANs, e.g., ARCNET and Ethernet, and provide the relay function described later in this section, are called *BACnet routers.* BACnet routers are "dynamic" in that they build, or update, their routing tables automatically using network layer protocol messages. Network layer protocol messages facilitate both the initialization and auto-configuration of dynamic routers and the flow of messages to, and between, routers.

By definition, a router is a device that is connected to at least two BACnet networks. Each attachment is through a *port* or physical network interface. A *routing table* consists of the following information for each port:

1. The MAC address of the port's connection to its network
2. The 2-octet network number of the directly connected network
3. A list of network numbers reachable through the port along with the MAC address of the next router on the path to each network number and the reachability status of each such network

4. In the case of routers capable of initiating a dial-up connection to another network, a telephone number

The *reachability status* is an implementation-dependent value that indicates whether the associated network is able to receive traffic via the next router on the directly connected network. The reachability status must be able to distinguish between "permanent" failures of a route, such as might result from the failure of a router, and "temporary" unreachability due to the imposition of a flow control restriction.

Some functions assigned to the network layer by the OSI Basic Reference Model are not required in BACnet. One of the most common involves selecting a communications path between source and destination machines based on an optimization algorithm. This is not required because BACnet internetworks may have at most a single, active path between any two devices, a constraint that greatly reduces the complexity of the network layer. Another common network layer function is message segmentation and reassembly. To obviate the need for these capabilities at the network layer, BACnet imposes a limitation of 507 octets for the length of messages passed to the data link's MAC layer, a message length that can be accommodated by all BACnet LANs. This guarantees that each message may be passed from router to router without regard for the underlying data link technologies that may be encountered along the path from source to destination. (Longer message lengths may be used locally if a particular LAN supports them.)

Figure 7.4 shows the data flow in the BACnet protocol stack. Just as a BACnet APDU is comprised of application layer protocol control information (APCI) and data (BACnet messages), so a BACnet NPDU consists of *N*PCI and data. The data passed to the network layer is called a *network layer service data unit* (NSDU) and, as Fig. 7.4 shows, consists solely of a BACnet APDU, encoded as described in the previous section.

As indicated above, the network layer protocol has to fulfill two objectives: to relay BACnet APDUs from one network to another and to provide the capabilities to maintain the routing tables used by BACnet routers. Let's look at how this layer works.

Each NPDU begins with a version number octet followed by a control octet that indicates the format of the subsequent network protocol control information (NPCI). All BACnet messages include, at a minimum, these two octets. Usage of the bits in the control octet is:

Bit 7: 0 NSDU contains BACnet APDU; 1 contains network layer message. Message type is in next octet.

Bit 6: 0 if no Vendor ID field; 1 if Vendor ID is present. (This field is used to allow proprietary network layer messages.)

Bit 5: 1 if DNET is present; 0 otherwise.

Bit 4: 1 if DLEN is present; 0 if DLEN is omitted (DADR is 1 octet long).

Bit 3: 1 if SNET is present; 0 otherwise.

Bit 2: 1 if SLEN is present; 0 if SLEN is omitted (SADR is 1 octet long).

Bits 1,0: Network priority where:

B′11′ = Life Safety message

B′10′ = Critical Equipment message

B′01′ = Urgent message

B′00′ = Normal message

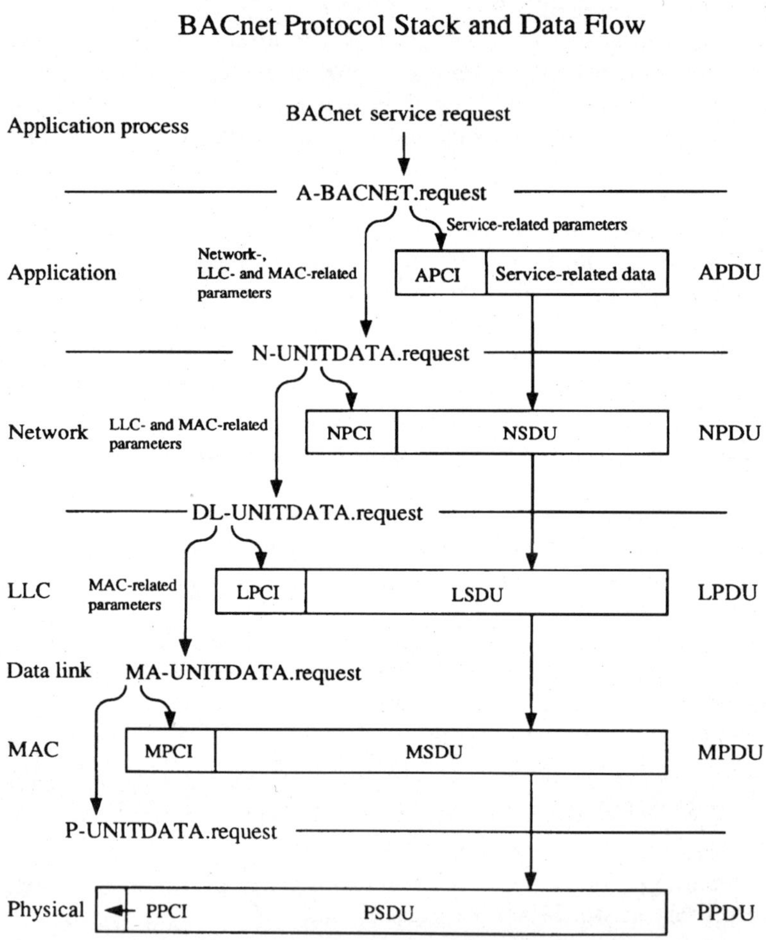

FIGURE 7.4. The BACnet collapsed architecture protocol stack. Each layer, or sublayer, adds its own protocol control information to the protocol data unit from the layer above. The figure shows that a BACnet implementation actually involves five distinct protocols, one for each layer or sublayer.

In the above:

DNET = 2-octet ultimate destination network number

DLEN = 1-octet length of ultimate destination MAC layer address (a value of 0 indicates a broadcast on the destination network, otherwise present only if the length of the destination MAC layer address is > 1)

DADR = Ultimate destination MAC layer address

SNET = 2-octet original source network number

SLEN = 1-octet length of original source MAC layer address (present only if length > 1)

SADR = Original source MAC layer address.

If Bit 7 of the control octet is 1, no BACnet APDU is present. Rather, the NPDU is conveying a message used strictly for the operation of the network layer.

Next Octet	Message
X'00'	Who-Is-Router-To-Network
X'01'	I-Am-Router-To-Network
X'02'	I-Could-Be-Router-To-Network
X'03'	Reject-Message-To-Network
X'04'	Router-Busy-To-Network
X'05'	Router-Available-To-Network
X'06'	Initialize-Routing-Table
X'07'	Initialize-Routing-Table-Ack
X'08'	Establish-Connection-To-Network
X'09'	Disconnect-Connection-To-Network
X'0A'-X'7F'	Reserved by ASHRAE
X'80'-X'FF'	Available for Vendor Proprietary messages

The I-Could-Be-Router-To-Network, Establish-Connection-To-Network, and Disconnect-Connection-To-Network messages relate to the operation of dial-up routers, i.e., routers that are only occasionally operational or that can connect to various networks depending on the establishment of a telephone conncetion.

The network layer relay algorithm is summarized in Fig. 7.5. The basic idea is that a router inspects each message addressed to it to see if the NPCI contains a DNET field; if it does, this signifies that the message needs to be sent to another network or, in the case where the global broadcast DNET is present, all networks. If the message originated on the network from which it arrived, the router adds the address of the originating node (SADR) to the NPCI along with its network number (SNET), thus relieving individual nodes from having to know which network they are on. The router then figures out (by looking in its routing table) whether the destination

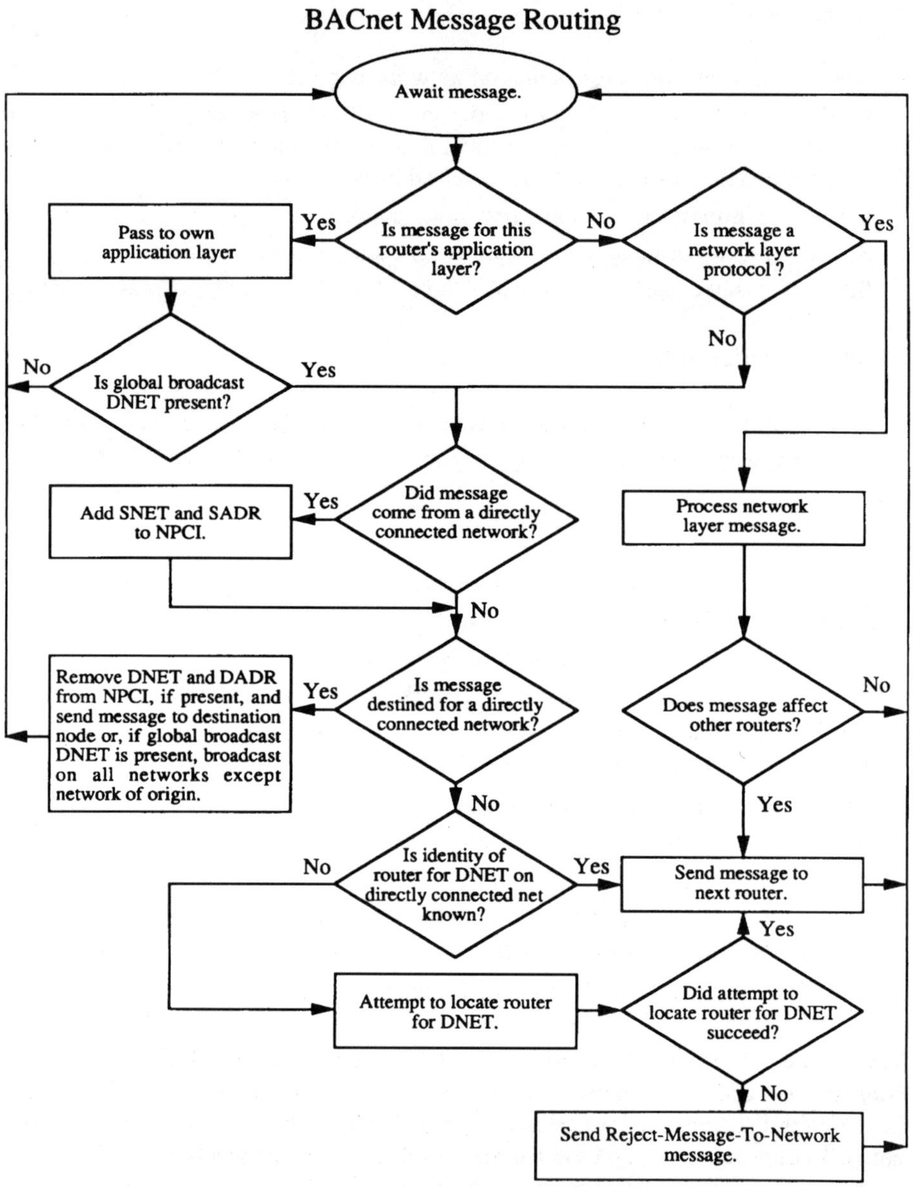

FIGURE 7.5. A schematic representation of the message processing procedure of BACnet routers. Note that BACnet routers may also perform DDC functions in addition to their routing duties.

network is directly connected or has to be reached via another router. In the former case, the destination network information (DADR, DNET) is stripped out of the message, and the message is sent to the destination node using its MAC address. In the latter case, the router looks up the MAC address of the next router on the path to the destination network and sends the message to it for subsequent relay.

The network layer messages cited above come into play if a router does not know where to send a message for relay. In this case it attempts to discover the identity of the next router on the path to a given destination network by broadcasting a Who-Is-Router-To-Network message containing the network number of the destination network. If a router on a directly connected network knows how to reach the indicated destination network, it broadcasts an I-Am-Router-To-Network message back to the router that broadcast the original query (and any other routing node that wants to update its tables). If the router cannot find a router to relay its message it sends a Reject-Message-To-Network message back to the originating node.

The Router-Busy-To-Network and Router-Available-To-Network messages are used by routers for flow control purposes, much like the ASCII XON/XOFF (X′11′/X′13′) characters are used for flow control between dumb terminals and computers.

The exact functioning of the messages used to control dial-up router operation is still under development as of late 1993.

7.5.1.1. *Use of BACnet with IP and IPX* Since many potential BACnet sites already have various networks in place, complete with routers, the BACnet committee decided to describe in appendices to the standard how BACnet messages could be conveyed using such existing facilities. The two most common networks make use of the Department of Defense's (DoD) Internet Protocols (IP) (described in the Defense Data Network Protocol Handbooks) and Novell's IPX (described in literature from Novell). Both protocol "suites" (sets of protocols) are capable of encapsulating/decapsulating BACnet messages and conveying them across an IP or IPX internet using a technique known as *tunneling*.. The appendices describe the procedures in terms of separate devices called BACnet/Internet Protocol Packet-Assembler-Disassemblers (B/IP PADs) in the case of IP and BACnet/IPX Packet-Assembler-Disassemblers (B/IPX PADs) in the case of IPX. It is also possible that a vendor might choose to build the necessary functionality into BACnet nodes directly. In this section we will take a brief look at how IP can be used. The appendix dealing with IPX is nearly identical.

Basically, a B/IP PAD is a device that implements both the BACnet network layer as described above in Section 7.5.1 as well as the DoD User Datagram Protocol (UDP) and Internet Protocol (IP). UDP is a simple, unacknowledged protocol that allows user-supplied blocks of data (datagrams) to be sent to any internet node by means of IP's routing capabilities.

Upon receipt of a BACnet packet from the local network, the B/IP PAD looks at the network layer information to see if a remote network number has been supplied. If so, the B/IP PAD then consults an internal table consisting of entries mapping network numbers, IP addresses of B/IP PADs, and the IP address of an IP router on the local network that represents the next hop for the IP datagram. If an appropriate entry is found, the B/IP PAD encapsulates the BACnet message in a UDP packet where the message represents the data portion of the packet. The B/IP PAD then sends the UDP packet to the local IP router indicating its own IP address as the source address and the IP address of the B/IP PAD corresponding to the remote network number contained in the BACnet message as the destination IP address. If the BACnet message is a global broadcast, the message is sent to all B/IP PADs contained in the table. Between the B/IP PADs standard IP procedures are followed.

Upon receipt of an UDP packet from an IP router, the B/IP PAD extracts the data portion of the UDP datagram and sends it as a BACnet message on the local network.

An implementation on a BACnet 8802-3 LAN might be configured as shown in Fig. 7.6. BACnet packets are differentiated from IP packets by different protocol identifiers in the header portion of the messages. In the case shown in the figure, each packet actually appears twice on the network, once as a BACnet message and once as an IP message. B/IP PADs could also be constructed to implement the IP routing procedure for any IP packet in addition to performing the BACnet encapsulation/decapsulation. Such a device could be called a B/IP PAD/Router. This is illustrated in Fig. 7.7.

7.5.2 BACnet LANs

The last piece of the BACnet puzzle is the question of how to get a message from one device to another. Unfortunately, there was no one single LAN that met all the criteria that had been established. As a result, the SPC chose to select or develop *four* LAN/data link technologies that, taken as a group, can probably meet any practical requirement. Two of the four, Ethernet, or more properly ISO 8802-3, and ARCNET have already been described. Both are already used by various DDC vendors and there is, therefore, an experience base in the industry which should help to bring BACnet-conformant devices to the marketplace a bit sooner than might otherwise be the case. Technically, the two LANs are likely to be nearly equivalent in terms of performance for the kind of messages most prevalent in integrated DDC systems. ARCNET tends to be less costly and, because it is a token-passing LAN, it is deterministic, i.e., you can predict and measure the worst-case time between network access opportunities for any given node. Ethernet, on the other hand, can be even faster than ARCNET for applications such as uploading and downloading, and equipment, such as repeaters and bridges, is available from a very wide selection of suppliers.

The other two BACnet data link layers, MS/TP and point-to-point, are described below.

7.5.2.1 MS/TP MS/TP is a master-slave/token-passing data link layer developed entirely within ASHRAE. Its development was necessitated by the desire to be able to use inexpensive EIA-485 hardware with BACnet and the absence of a standard serial asynchronous data link layer protocol. Since almost every manufacturer of

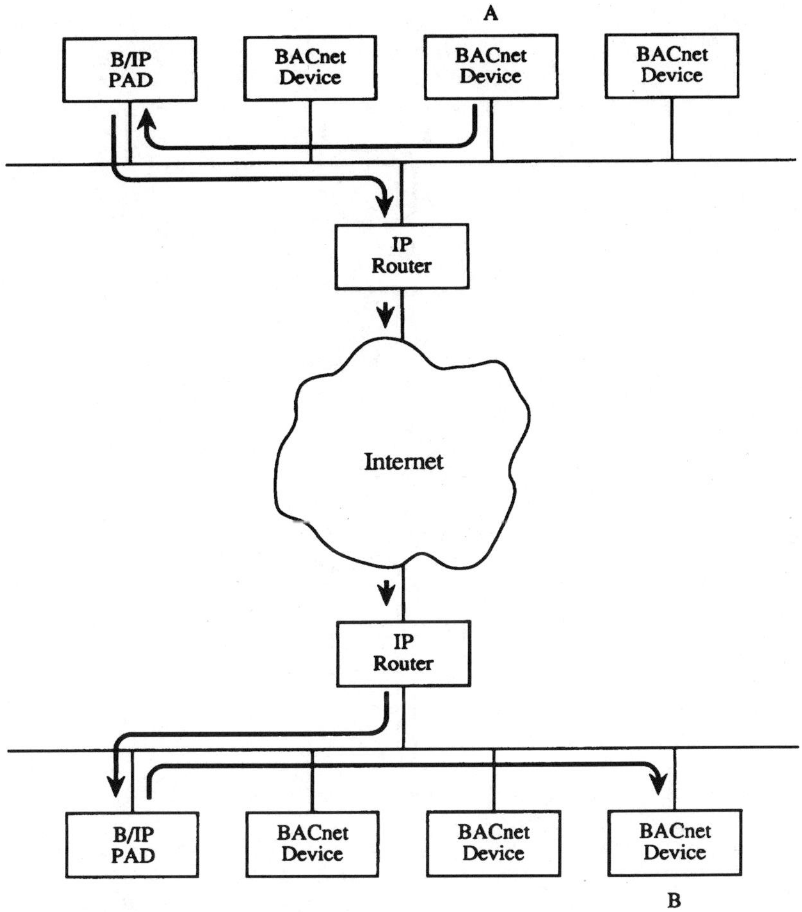

FIGURE 7.6 Two BACnet devices communicating via B/IP PADs and standard IP routers. Notice that each message appears on the network twice, once as a BACnet message, once as an UDP datagram.

DDC equipment has some form of proprietary EIA-485 network for his own use, it seemed reasonable that the SPC could, with all the experience represented on the committee, "roll its own."

While ARCNET and Ethernet are both strictly peer-to-peer LANs, MS/TP envisions the possibility of both master nodes, which are peers of one another, and slave nodes, which are subservient to all the master nodes. The main difference between masters and slaves is that the slaves are never in possession of the MS/TP token; also, they don't have to be smart enough to deal with the case of token loss and regeneration.

Here is an overview of MS/TP. MS/TP uses a token to control access to an EIA-485 bus network. A master node can transmit a data frame only when it holds

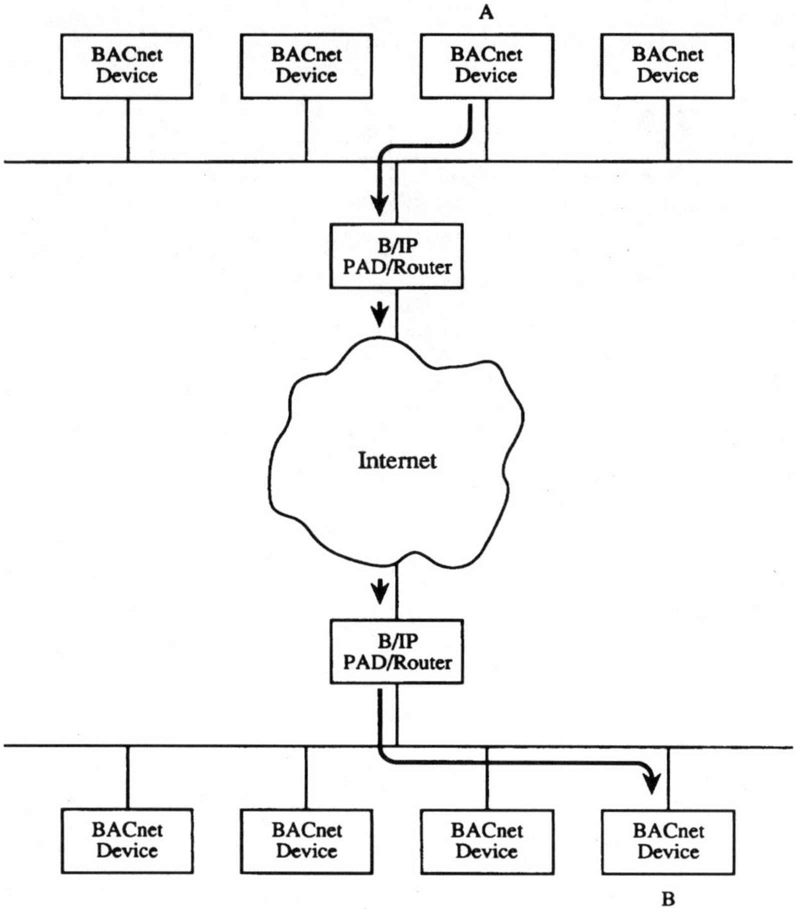

FIGURE 7.7. Two BACnet devices communicating via combined B/IP PAD/Routers. In terms of communication traffic this would be the preferable approach but would require the IP router itself to understand BACnet. This would eliminate the beneficial possibility of using "off-the-shelf" IP routers, at least on the local networks.

the token. A slave node can transmit only in response to a request from a master node. After sending up to a configurable maximum number of frames (and waiting for any expected replies from the slaves), a master node must pass the token to the next master node.

Token frames are not acknowledged. If the acknowledgment of a token were lost, the token's sender might retry the token pass, resulting in the creation of two tokens (a bad situation!). Instead, after a node passes the token, it listens to see if the intended receiver node begins using the token. Usage is defined as the reception of at least one octet more than 3 but less than 10 octet times after the final octet of the token frame is transmitted. (An octet time is the time it takes to send a data octet, including a start and stop bit, at whatever baud rate is in use on the MS/TP LAN.)

A conventional token bus network such as ARCNET does not distinguish between requests and replies: both are passed in the same type of frames, which are sent only when a node has the token. Since MS/TP defines slave nodes which never hold the token, a means must be provided to allow replies to be passed from slave devices.

When a BACnet-confirmed service request is sent to an MS/TP *master* node, the reply will be returned at some later time when the destination node holds the token. When a confirmed service request is sent to an MS/TP *slave* node, the sender waits for the reply to be returned from the slave before passing the token.

All MS/TP frames (messages) use a common format:

(Idle)	At least 3 octet times of idle (binary 1)
Preamble	Two octet preamble: X′55FF′
Frame type	One octet (see below)
Destination address	One octet (0-254, 255=Broadcast)
Source address	One octet (0-254)
Length	One or two octets, depending on length
Data	(Optional, depending on Frame type)
CRC	Two octets, CRC-CCITT calculation
(Pad)	(Optional, at most one X′FF′ of padding).

The preamble sequence X′55FF′ provides a sequence not likely to arise from random noise and that is easily distinguished on a oscilloscope or other data display. It is used as a second level of assurance (after the idle) against improper framing.

```
111..11111  0101010101  0111111111  0XXXXXXX etc
 ←idle→       ←55→        ←FF→        ←frame→
```

Note that each nonidle octet has a single 0 start bit, a 1 stop bit, and 8 data bits transmitted least-significant bit first.

The MS/TP frame types and their use are summarized next.

Frame Type 00: Token The Token frame is used to pass network mastership to the next master on the logical token ring.

Frame Type 01: Poll For Master The Poll For Master frame is transmitted by Master nodes during configuration, and periodically during normal network operation. It is used to discover the presence of other master nodes on the network and to determine a successor node in the token ring.

There are no data octets in Poll For Master frames, and the length octet is zero.

Frame Type 02: Reply To Poll For Master This frame is transmitted as a reply to the Poll For Master frame. It is used to indicate that the node sending the frame wishes to enter the token ring.

Frame Type 03: Test_Request This frame is used to initiate a loopback test of the MS/TP to MS/TP transmission path. The length of the data portion of a Test_Request frame may range from 0 to 510 octets.

Frame Type 04: Test_Response This frame is used to reply to Test_Request Frames. The length of the data portion of a Test_Response frame may range from 0 to 510 octets. The data, if present, matches what was sent in the initiating Test_Request.

Frame Type 05: BACnet Data Expecting Reply This is the format of frames which carry BACnet confirmed service requests. The length of the data portion of a BACnet Data Expecting Reply frame may range from 0 to 510 octets.

Frame Type 06: BACnet Data Not Expecting Reply This is the format of all frames which carry BACnet unconfirmed service requests, or BACnet replies to confirmed services. The length of the data portion of a BACnet Data Not Expecting Reply frame may range from 0 to 510 octets.

The detailed operation of master and slave nodes is defined in the standard through the use of *finite-state machine* representations, one for master nodes and one for slaves, and several state variables and timers. These should be consulted by anyone interested in a more complete description of this part of BACnet.

BACnet also describes the operation of MS/TP repeaters. Repeaters are to be used if more than 32 MS/TP nodes need to be interconnected or the physical extent of the network needs to be greater than 4000 ft.

7.5.2.2 Point-to-Point The BACnet Point-to-Point (PTP) data link layer protocol (formerly called Dialup) defines mechanisms by which two BACnet devices can communicate using serial asynchronous point-to-point communication mechanisms such as typified by dial-up telephone connections. These mechanisms are accessed through the use of the EIA-232 signaling standard and generally involve modems or line drivers. BACnet does not attempt to specify the means by which the physical connection is established, or whether the modems themselves may implement error detection or error correction. The connection may be permanent or temporary, the latter case appropriate to a true dial-up connection. Rather, it specifies a protocol which allows two BACnet network layer entities to establish a BACnet PTP connection, reliably exchange BACnet PDUs, and perform an orderly termination of the connection using a physical connection already established by other means. The "other means" could be, for example, a human being or an autodialer.

A PTP connection can differ from a local area network connection in several ways: it requires a much simpler medium access control mechanism; it is usually temporary in nature; it can be significantly slower; and it is capable of full-duplex operation.

The BACnet PTP protocol does not assume that the answering device is necessarily in a state where it can accept BACnet PDUs containing binary information. The same serial device port that is dialed into by a BACnet device may also be dialed into by a human operator using a simple ASCII terminal. Therefore, the BACnet connection initialization procedure (as opposed to the *physical* connection procedure) is performed entirely using ASCII printable character sequences.

The connection process also provides optional password protection for the case where privileged knowledge of a phone number is not considered enough security. The configuring and checking of the password parameter is left up to individual implementors.

The PTP protocol is divided into three phases: data link establishment, data exchange, and data link termination.

Data Link Establishment Once a physical connection has been established between the calling device and the answering device, the calling device transmits an ASCII *trigger sequence* consisting of the letters BACNET. If the answering device understands the significance of this character string, it sends back a Connect Request frame containing its device identifier and, optionally, a password. The calling device then verifies the device ID (and, possibly, password) and sends back a Connect Response frame with its ID and password. If both the caller and answerer are happy with this information, Heartbeat (XON) frames are exchanged (described in more detail below) to indicate each side's willingness to receive data request frames. The connection is then deemed to be established.

Data Exchange At this point, the two devices can begin to exchange BACnet (binary format) PDUs bi-directionally. The BACnet PDUs are imbedded alternately in Data Request 0 and Data Request 1 frames which are acknowledged using Data Response 0 and Data Response 1 frames, respectively. This alternating sequence number approach is similar to ISO 8802-2 Acknowledged Connectionless LLC. To ensure activity on the data link in the absence of BACnet messages, each device transmits Heartbeat frames at an interval approximately one-third to one-quarter that of the inactivity timer. Both the Heartbeat and Data Response frames carry flow control information that indicates whether the originator of the frame is ready to receive additional frames. To allow for the possibility of modems that might inappropriately respond to the ASCII XON or XOFF flow control characters (X′11′ or X′13′) by ceasing to transmit, these characters are replaced in the binary data stream by X′91′ and X′93′, respectively, and preceded by a *data link escape* character of X′10′.

Data Link Termination Either the calling device or the answering device may terminate the connection using Disconnect Request and Disconnect Response frames. The connection remains in effect until a request for termination has been accepted by the peer device, one of the devices determines that the physical layer connection has been lost, or until a local timer expires indicating that the peer device is no longer active.

PTP Frame Types All PTP frames have the same format:

Preamble	2-octet preamble: X'55FF'
Frame type	1 octet (see below)
Length	2 octets, most significant octet first
Header CRC	1 octet
Data	0 to 2048 octets (varies with frame type)
CRC	2 octets, CRC-CCITT calculation (if data present)

The PTP frame types and their use is summarized next.

Frame Type 0: Heartbeat (XOFF) A frame of this type is transmitted periodically when no other data is ready for transmission to indicate to the peer device that the data link is still active. This frame also indicates that the local device is *not* ready to receive further data request frames.

Frame Type 1: Heartbeat (XON) This frame is used as above but also indicates that the local device *is* ready to receive further data request frames.

Frame Type 2: Data Request 0 The Data Request 0 frame is used to convey data (NPDUs) with a sequence number of 0. The length of the data portion of the frame may range from 0 to 2048 octets.

Frame Type 3: Data Request 1 The Data Request 1 frame is used to convey data (NPDUs) with a sequence number of 1. Again, the length of the data portion of the frame may range from 0 to 2048 octets.

Frame Type 4: Data Response 0 (XOFF) The Data Response 0 frame is used to acknowledge a correctly received Data Request 0 frame. The data field is omitted. The use of this frame also signals that the sending device is not able to receive further data request frames.

Frame Type 5: Data Response 1 (XOFF) The Data Response 0 frame is used to acknowledge a correctly received Data Request 0 frame. The data field is omitted. As above, the use of this frame also signals that the sending device is not able to receive further data request frames.

Frame Type 6: Data Response 0 (XON) The Data Response 0 frame is used to acknowledge a correctly received Data Request 0 frame. The data field is omitted. The use of this frame also signals that the sending device is able to receive further data request frames.

Frame Type 7: Data Response 1 (XON) The Data Response 0 frame is used to acknowledge a correctly received Data Request 0 frame. The data field is omitted. As above, the use of this frame also signals that the sending device is able to receive further data request frames.

Frame Type 8: Data Reject 0 This frame is used to reject a Data Request 0 in which the header segment of the frame was correctly received, but the data portion of the frame contained errors.

Frame Type 9: Data Reject 1 As above, for a Data Request 1 frame.

Frame Type 10: Connect Request Issued by the answering device after receipt of the string BACNET.

Frame Type 11: Connect Response Issued by the caller after receiving a Connect Response frame.

Frame Type 12: Disconnect Request The Disconnect Request frame may be issued by either the calling or answering device when it wishes to discontinue the BACnet PTP dialog. The data portion of the frame conveys the reason for requesting a disconnect and is one octet in length. The possible values for the data field are:

X′00′ No more data needs to be transmitted
X′01′ The peer process is being preempted
X′02′ The received device identifier is invalid
X′03′ The received password is invalid
X′04′ Other

Frame Type 13: Disconnect Response The Disconnect Response frame is used to indicate that the responding device has accepted the request to disconnect.

The detailed rules for the operation of the PTP protocol are again described in the BACnet specification using a finite-state machine model and should be consulted for further information.

7.6 EXTENSIBILITY ISSUES

A major concern of all standards bodies working with rapidly changing technologies is obsolescence. How can standards be developed that will meet current needs but not hamper technological innovation and change? In the case of BACnet, considerable attention was paid to this critical issue since the SPC realized that acceptance of the standard could well depend on the standard's ability to adapt to changing times. In essence, the question is how can nonstandardized information be transferred within the constraints of the BACnet standard? BACnet addresses this concern by providing three distinct mechanisms for such communication.

7.6.1 The PrivateTransfer Services

The PrivateTransfer services, discussed in Section 7.3.5, are used to transfer blocks of information of unspecified content and are therefore used to transfer nonstandard messages. Each service request consists of a vendor ID parameter, a service number, and zero or more unspecified parameters. If a response is required, the confirmed version of the PrivateTransfer service could be used which contains a Result Block parameter, again of unspecified content.

These services could be used, for example, to embed existing proprietary protocols within BACnet PDUs or to provide for other nonstandard capabilities.

7.6.2 Adding Proprietary Properties to a Standardized Object

BACnet allows a vendor to add proprietary properties to extend the capabilities of a standard object. Proprietary properties receive the same support from BACnet services as standard properties and therefore can be accessed and manipulated in a manner identical to standard properties.

If a proprietary property is to be a commandable property, additional properties which fulfill the role of the standard Priority_Array and Relinquish_Default properties must also be provided in order to accommodate the priority synchronization mechanism.

Vendors may add proprietary properties to a standard object by modifying the object definition within a device. Proprietary properties are enumerated with PropertyIdentifier values greater than 511. These PropertyIdentifiers can be used in any BACnet service which uses a PropertyIdentifier as an argument.

Other enumerations within BACnet may be extended in the same way.

7.6.3 Adding Proprietary Objects to BACnet

To accommodate applications where the defined set of standardized objects is not adequate, BACnet allows a vendor to add proprietary objects. To enhance extensibility, BACnet provides the same support for proprietary objects as for standard objects.

Vendors may add proprietary objects to BACnet by extending the object type enumeration to include new nonstandard objects. Proprietary objects have enumerated values greater than 63. These object types can be used in any BACnet service which uses an object type as an argument. The properties of these objects may include both standard and proprietary data types.

With all of the standard objects and services, plus the mechanisms described in this section, BACnet should be capable of conveying virtually any type of information with a reasonable degree of efficiency. But if it can't there is always Version 2!

7.7 CONFORMANCE AND SPECIFICATION

While most people think it is a great idea to have data communication standards, the standards are often bedeviled by the twin bugaboos of conformance and specification! In the absence of clear conformance criteria, anyone who has attempted to implement a standard could claim that his products conform to it. In the absence of clear specification guidelines, designers would have little choice but to specify that products "shall conform to the standard." In either case, the purchaser of the products is likely to have difficulties.

Examples of this problem abound. Have you ever tried to hook up an EIA-232-compatible printer or modem? People have made small fortunes writing treatises on how to interconnect "standard" EIA-232 devices! Another infamous example is the Tri-Service EMCS Specification of the early eighties that was prepared by the Navy to provide guidelines for control system designers in the Department of Defense. It was eventually (ab)used by much of the federal government. The problem was not with the document itself: it was a comprehensive attempt to lay out the requirements for "small," "medium," and "large" EMCSs and contained many excellent provisions. The problem was that people took the guide spec, slapped their agency cover pages on it, and put it out on the street saying, in effect, "give us one of these," a completely meaningless request. Sadly, some vendors claimed they could! The result was a debacle that gave the digital controls industry such a black eye that it is only now recovering from it.

The issue before the BACnet committee was therefore how to avoid the type of problem encountered a decade earlier by the Tri-Services folks. What does it mean to "conform" to BACnet and how does a building automation system designer go about specifying BACnet in a reasonable way without having to be a computer guru?

Let's first look at what it means for a device to conform to BACnet. There are three requirements: (1) the manufacturer must supply a protocol implementation conformance statement (PICS), a written document identifying the particular BACnet objects and services that are implemented in the device; (2) the device must pass a conformance test which verifies the correct implementation of the standard object types and services indicated in the PICS; and (3) the device must support one of the four data link layer options specified in the standard and one of the physical layers compatible with that data link layer. Let's examine each of these provisions in more detail.

First, the PICS. A BACnet PICS has five sections. The first provides some basic information to identify and describe the BACnet device; the second enumerates the standard object types that are supported and indicates for each object type whether only the standard properties are supported or if the optional properties are also supported; the third enumerates the BACnet application services that are supported and for each service indicates if the implementation can initiate the service request, respond to the service request, or both; the fourth enumerates the BACnet functional groups that are supported (see below); and the last section indicates the data link and physical layer options that are implemented. Armed with a PICS, designers should be able to compare apples to apples!

The second requirement is the conformance test. This is seen by many as the most critical component of the three. The test consists of a collection of communication dialogs drawn from a standard test suite in such a way as to test each object type and service for which support is claimed in the PICS. At this moment, however, it has yet to be determined who is going to conduct the tests or exactly what the tests will be. This will hopefully be resolved shortly.

The third conformance requirement is that the device must support at least one of the protocols referenced or defined in the standard, i.e., ARCNET, 8802-3, MS/TP, or PTP. Devices that do not may claim that they "conform to BACnet with a nonstandard data link layer" if they meet the PICS and conformance test requirements. This would allow someone to build a BACnet network using, for example, FDDI, microwave, or powerline carrier technology. As long as the devices involved meet the other conformance requirements, the nonconformant LAN could be bridged to BACnet networks using the standard data links.

To aid specifiers, the provisions of BACnet that relate to objects and services have been arranged in categories referred to as *conformance classes* and *functional groups.*

The conformance classes are numbered 1 to 6 and are hierarchical in nature. The requirements for each class include all of the requirements of all of the other classes with a lower number. Each class consists of one or more services that must be supported to claim membership in the particular class. For each service a distinction is made between a requirement to be able to *initiate* the service request and a requirement to be able to *execute* a service request of the specified type. Certain standard object types may also have to be supported to meet the requirements of a particular conformance class. Conformance class 1, for example, requires that a device must be able to execute a ReadProperty request and must possess a Device object. Conformance class 2 adds the requirement that a device must be able to execute a WriteProperty request.

Although conformance classes can be useful to confirm that a certain core group of communication capabilities is present, functional groups are an attempt to facilitate the specification of capabilities by operational function. An enumeration of the nine functional groups should make this clear:

- *Clock.* Provides the capabilities generally associated with a clock. Support for this functional group requires being able to execute the TimeSynchronization service and supporting the Local_Time and Local_Date properties of the Device object.
- *Hand-held Workstation.* Provides the capabilities generally associated with handheld workstations.
- *Personal Computer Workstation.* Provides the capabilities generally associated with a personal computer work station. These include *all* the object access services and file services.
- *Event Initiation.* Provides the capability to define alarms and events, detect the occurrence of such alarms or events, and initiate notification messages to subscribing devices.
- *Event Response.* Provides the capability to respond to a notification that an event has occurred, including the issuance of an AcknowledgeAlarm service request by a human operator.
- *Files.* Provides the capability to read, write, upload, and download files.
- *Reinitialize.* Represents the capability to reinitialize a device from a remote location by supporting execution of the ReinitializeDevice service request.

- *Virtual Operator Interface.* Provides the capabilities needed for the operator side of a virtual terminal session.
- *Virtual Terminal.* Provides the capabilities needed for the server side of a virtual terminal session.

To summarize, the underlying idea behind functional groups is to provide a shorthand way for people to refer to the portions of the BACnet protocol that are needed to perform particular functions without having to know all of the details. Devices can be expected to interoperate with respect to a given function so long as they are in the same conformance class and support the same functional groups. If devices are in different conformance classes, they should be expected to interoperate only with respect to the functionality of the lower conformance class and any common functional groups that are supported by both devices. Many well-designed networks will undoubtedly contain devices representing a mixture of conformance classes and functional groups.

A final warning: this chapter describes BACnet as it was at the time of writing, mid 1993. The standard is currently in public review draft form and can be expected to change, at least in terms of some of the details, before it is finally published as a final standard. If you are interested in the most recent and authoritative version, you should arrange to get a copy from ASHRAE at the address listed in the references section.

REFERENCES

ASHRAE Standard 135P, First Public Review Draft, *BACnet: A Data Communication Protocol for Building Automation and Control Networks*, Atlanta, GA: ASHRAE, August 1991.

DDN (Defense Data Network) Protocol Handbook, Volumes 1–3, NIC 50004, 50005, and 50006.

IEEE 754-1985, *IEEE Standard for Binary Floating-Point Arithmetic,* New York: IEEE, 1985.

ISO 8824, *Information Technology—Open Systems Interconnection—Specification of Abstract Syntax Notation One (ASN.1)*, Geneva: ISO, 1990.

ISO 8825, *Information Technology—Open Systems Interconnection—Specification of Basic Encoding Rules for Abstract Syntax Notation One (ASN.1)*, Geneva: ISO, 1990.

SOURCES FOR STANDARDS

ASHRAE: ASHRAE, 1791 Tullie Circle, N.E., Atlanta, GA 30329.

DoD: DDN Network Information Center, SRI International, 333 Ravenswood Avenue, Room EJ291, Menlo Park, CA 94025.

ISO: American National Standards Institute, 11 West 42nd Street, 13th Floor, New York, NY 10036.

Chapter 8

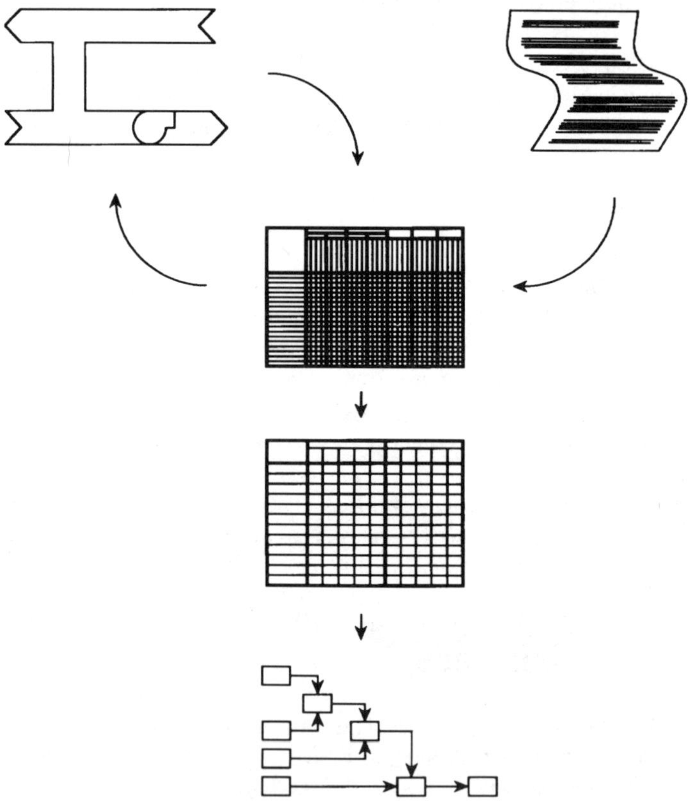

DDC System Design: A Case Study

SUMMARY

In this chapter we will design a complete direct digital control system using a methodology that both simplifies the design task and results in high quality documentation, an aspect of DDC jobs that has often been inadequate. In fact, the lack of documentation standards has been one of the main impediments to the universal acceptance of DDC as the control technique of choice for building HVAC systems.

The technique we will use was first proposed by my colleague Clay G. Nesler, currently Manager of Advanced Systems Development at Johnson Controls, in conjunction with the ASHRAE Professional Development Seminar on DDC that we have been teaching together since the spring of 1990. We have been trying jointly to refine and streamline it since its introduction.

For our case study we will design the controls for a single-zone, constant-volume air handler. As we work through the example, it should become clear how straightforward it would be to design and document a more complicated system, perhaps involving humidity control and variable air volume. The components of our "top-down" design method are shown schematically in Fig. 8.1.

In the discussion that follows we will examine each part of the method in detail.

8.1 SYSTEM DIAGRAM

For the purposes of this example, let's assume that the HVAC engineer has already designed the air handler in terms of laying out the ductwork, appropriately sizing the fan and heating and cooling coils, and selecting the proper dampers, damper actuators, and motor contactor. From this design we can construct a system diagram as shown in Fig. 8.2A. The designations DA and VA stand for damper and valve actuator, respectively, C is for electrical contactor, and H/C and C/C represent the heating and cooling coils. We will return to the diagram to add sensors and interface components after we study the sequence of operation and begin to construct the I/O

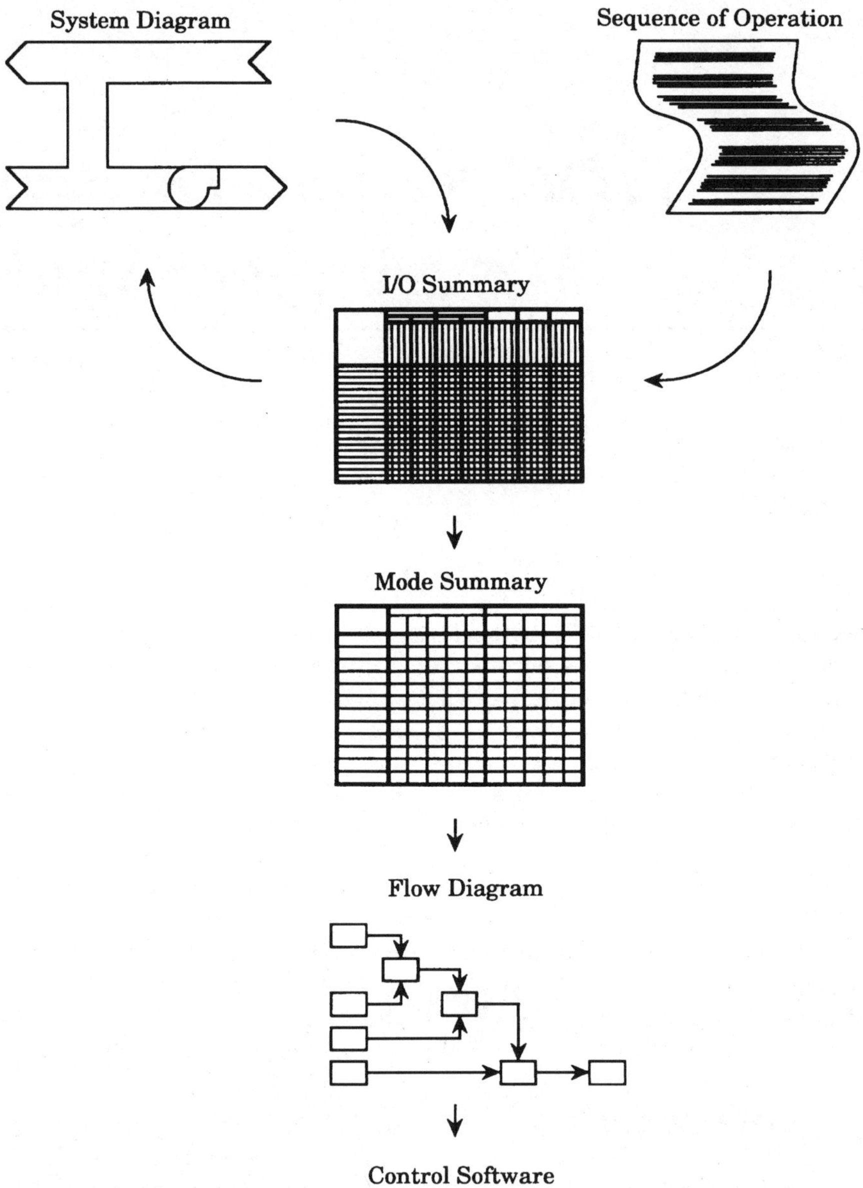

FIGURE 8.1. The components of the top-down DDC design methodology described in the text. The technique both simplifies the design task and produces documentation suitable for both hardware and software implementation.

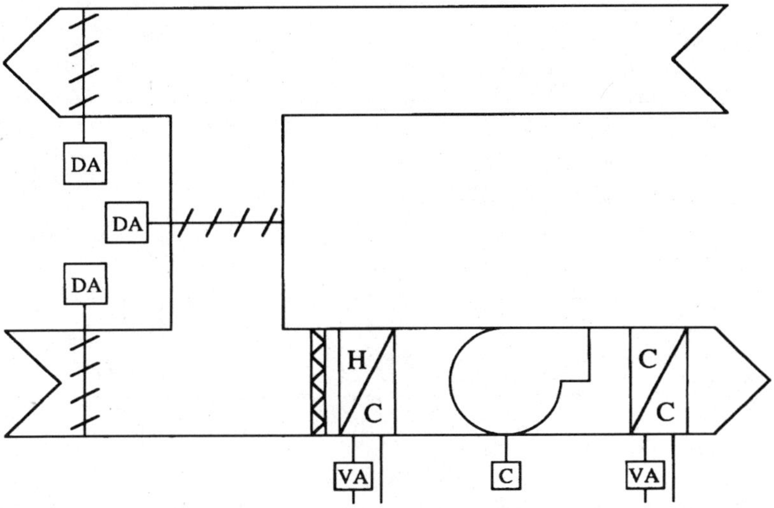

FIGURE 8.2A. A schematic of the basic single-zone, constant-volume air handler used in our example.

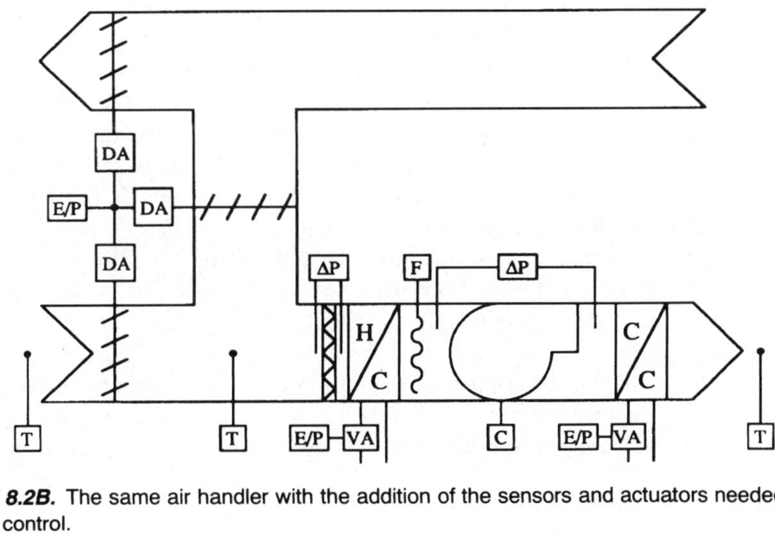

FIGURE 8.2B. The same air handler with the addition of the sensors and actuators needed for DDC control.

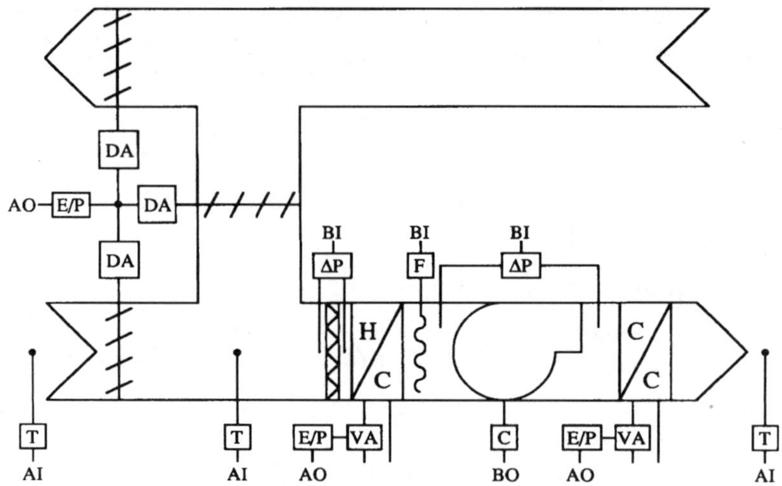

FIGURE 8.2C. The final installation showing the connections to the DDC input and output circuits.

(input/output) summary. This is the reason we have shown arrows leading both to and from the system diagram in Fig. 8.1.

8.2 SEQUENCE OF OPERATION

The sequence of operation verbally describes how the air handler is to be operated and controlled. It contains detailed descriptions of each mode of operation, control strategy, and alarm condition, including setpoints, limit values, temperature setback schedules, and any other relevant operating parameters.

A *mode of operation* is a description of how all of the various components of the air handler as a whole will be operated given a particular set of input conditions. For example, when the zone of the building served by the air handler is "occupied," i.e., the date and time fall within a certain schedule, the system is said to be in the *occupied mode*. In this mode, the fan is started and the heating and cooling valves and dampers are modulated so as to maintain the setpoint temperature in the zone, the "normal" operating condition. Note that the modes described in the sequence of operation are mutually exclusive. The air handler can only be in one mode at a time.

Control strategies, in contrast, describe how specific subsystems are to be controlled. Unlike the operating modes, several, if not all, of the control strategies may be executing concurrently.

Here is the sequence of operation for our single zone, constant volume air handler:

Sequence of Operation

1. *Occupied Mode*. When the date and time fall within the specified schedule, the system shall enter the occupied mode. The supply fan shall be turned on; the normally open heating and normally closed cooling valves and air dampers

shall be controlled by a sequenced proportional plus integral (PI) controller to maintain the room air temperature setpoint of 70°F.

2. *Unoccupied Mode.* When the date and time fall outside of the specified schedule and the room air temperature exceeds 55°F (or 60°F if the room air temperature has previously been less than 55°F), the system shall enter the unoccupied mode. The supply fan shall be turned off; the heating valve shall be set to fully open and the cooling valve and outside air dampers shall be set to fully closed.
3. *Setback Mode.* When the date and time fall outside of the specified schedule and the room air temperature is less than or equal to 55°F, the system shall enter setback mode. The supply fan shall be turned on; the heating valve shall be set to fully open and the cooling valve and outside air dampers shall be set to fully closed. Once in this mode, the system shall not leave it until the room air temperature exceeds 60°F.
4. *Fan Alarm Mode.* If the supply fan has been commanded to start and the supply fan status, after a delay of 30 s, indicates that the fan is not operating, the system shall enter fan alarm mode (ON). A stop command shall be sent to the supply fan; the heating valve shall be set to fully open and the cooling valve and outside air dampers shall be set to fully closed.

   If the supply fan has been commanded to stop and the supply fan status, after a delay of 30 s, indicates that the fan is continuing to run, the system shall enter fan alarm mode (OFF). A stop command shall be sent to the supply fan and the heating valve, cooling valve, and outside air dampers shall continue to be controlled normally as in the occupied mode.
5. *Room Temperature Control.* During the occupied mode, room temperature will be maintained by a PI control algorithm which will sequentially modulate the heating valve, outside and return air dampers, and cooling valve. As the PI controller output goes from 0 to 33%, the heating valve shall go from fully open to fully closed. As the PI controller output ranges from 34 to 66%, the outside air dampers shall go from a minimum position of 20% to fully open. As the PI controller output goes from 67 to 100%, the cooling valve shall go from fully closed to fully open.
6. *Dry Bulb Economizer Control.* When the system is in occupied mode and the outside air temperature rises above 65°F, the dry bulb economizer setpoint, the outside air damper will be set to a constant position of 20%, subject only to mixed air low control. Dry bulb economizer control shall have priority over room temperature control.
7. *Mixed Air Low Limit Control.* When the system is in the occupied mode, a proportional (P) control algorithm shall modulate the outside air dampers from 100 to 0% as the mixed air temperature drops from 40 to 35°F. The output of the controller shall be passed to a low select control block, which will adjust the outside air dampers to the lesser of the positions called for by this controller and the room temperature local loop controller. Mixed air low-limit control shall have priority over both room temperature and dry bulb economizer control.

8. *Fan Alarm.* Whenever the system is in fan alarm mode, the fan alarm shall be initiated.
9. *Filter Alarm.* If the differential pressure across the air filter should exceed 0.2 in. WC, the filter alarm shall be initiated.
10. *Freeze Alarm.* If the heating coil discharge temperature should fall below 35°F, the freezestat contacts shall close and the freeze alarm shall be initiated.

Armed with the sequence of operation and the system diagram, we can now determine what inputs and outputs we will need to implement the desired control.

8.3 I/O SUMMARY

Table 8.1 shows the completed I/O summary. Across the top are column labels that divide the table into five main sections: inputs, outputs, modes, strategies, and alarms. The modes, strategies, and alarms can be filled in just by going down through the sequence of operation and recording the paragraph titles. The sequence is divided into 10 paragraphs and there are the same number of labeled columns in the right half of the table. The list of inputs and outputs is compiled and entered in the rows of the table by going through each paragraph in turn and figuring out what sensors and actuators are needed to actually carry out each of the specified functions.

Let's start at the top of the sequence of operation with occupied mode. We need to be able to start and stop the fan through the DDC so we will need a starter relay or contactor that takes the DDC's low-power, low-current binary output signal (typically 5 or 10 VDC) and interfaces to the motor starter. On the left of the table we enter Supply Fan, enter a Start/Stop Contactor in the binary output column, and put an X in that column and one in the Occupied column. Since we are ultimately trying to control the room air temperature we will need a Room Air Temperature sensor. Let's make it an RTD sensor and start an RTD column under the analog input type label. X's go in the RTD and Occupied columns.

The description of the occupied mode also mentions controlling the valves and dampers. We will actuate these using pneumatic power interfaced to the computer through E/P transducers. Accordingly, let's add an E/P Transducer analog output type column and appropriate X's. Going through the remainder of the sequence of operation is just as easy.

At this point, we can return to the system diagram and draw in the components we have just entered in our I/O summary. See Fig. 8.2B. The great thing about Table 8.1 is that you can now read off at a glance how many of each type of sensor or actuator interface you will need to complete the job and for what functions they are going to be used.

8.4 MODE SUMMARY

Now let's build a mode summary table. See Table 8.2. Across the top are column labels for each input condition and output action. In the first column we list each of the modes of operation. Note that the fan alarm mode requires two rows in the table

TABLE 8.1 I/O Summary

	Input Type			Output Type		Modes of Operation					Control Strategies			Alarms		
	Analog	Binary		Analog	Binary											
Inputs/Outputs	RTD	Freezestat	ΔP Switch	E/P Transducer	Start/Stop Contactor	Occupied	Unoccupied	Setback	Fan Alarm		Room Temperature (PI)	Dry Bulb Economizer	Mixed Air Low Limit (P)	Fan	Filter	Freeze
Supply Fan					X	X	X	X	X							
Room Air Temperature	X					X	X	X		X						
Heating Valve				X		X	X	X	X		X					
Cooling Valve				X		X	X	X	X		X					
Dampers				X		X	X	X	X		X	X	X			
Outside Air Temperature	X								X							
Mixed Air Temperature	X									X						
Fan Status			X						X					X		
Filter Status			X						X						X	
Freeze Lockout			X							X						X

since we want to take a different action if we are trying to start the fan and the status feedback tells us that the fan failed to start as opposed to the case where we are trying to stop the fan and the feedback tells us that the fan is continuing to run. The value of this table is that it summarizes very compactly how each mode is entered and what is then supposed to happen.

8.5 FLOW DIAGRAM

The next to last step in the design process is to put together a flow diagram. A flow diagram is a graphical, schematic representation of the entire control system that depicts each sensor and output device and the logic, to be implemented in software in the DDC, that interconnects them. See Fig. 8.3.

Here are the general rules for creating the diagram. Start by arranging the inputs on the left and the outputs on the right. Analog logic signals are shown using solid lines while binary signals are dashed lines. In any given logic chain, the priority of the function blocks increases to the right. Thus, in the logic chain that determines the signal to be sent to the damper actuators, the override logic takes precedence over the mixed air low limit logic, which in turn takes precedence over the economizer switchover, which supercedes the minimum ventilation logic, which is in series with the output of the PI controller.

As to the depiction of the function blocks themselves, I'm sorry to have to relate that no standards currently exist in the building control industry. This is in contrast, again, to the situation in the industrial process control industry where there are actually *two* standards. The first is ISA-S5.1, *Instrumentation Symbols and Identification* and the second is the Scientific Apparatus Makers Association Process Measurement and Control standard PMC 22.1, *Functional Diagramming of Instrument and Control Systems*. Both of these standards provide a comprehensive set of symbols capable of representing virtually any control logic you could think of. This would be a great area for an ASHRAE standard. Any volunteers?

In any event, the logic blocks shown in Fig. 8.3 are really a melange of symbols from various sources. The operation of the blocks with binary outputs is conveyed by stating the relationship of the inputs that makes the output TRUE or ON, either under the block or on the output signal line. The purpose of the blocks with analog outputs is described within the block and, in some cases, with the range of output values to the right of the block.

If you prefer, the binary blocks could be made more generic using the Boolean logic description of the function. For example, the Start block is really an OR function; the Override block is also an OR with the Fan Status input inverted, i.e., the input to the OR is FALSE or OFF if the Fan Status input is TRUE or ON and vice versa. The Alarm block is an Exclusive OR with a delay timer built in. The Stop block is an AND function with the Alarm input inverted. The Setback block is a comparator. The blocks labeled HSEL and LSEL select and pass through the higher or lower of their inputs, respectively. The Switch blocks pass the signal value shown on the top of the block when the binary input value is ON.

TABLE 8.2 Mode Summary

	Input Condition				Output Action			
Mode of Operation	Schedule	Room Air Temp.	Supply Fan S/S	Supply Fan Status	Supply Fan	Heating Valve	Cooling Valve	Outdoor Air Dampers
Occupied	Occupied				On	Normal	Normal	Normal
Unoccupied	Unoccupied	>55°F			Off	Fully open	Fully closed	Fully closed
Setback	Unoccupied	≤55°F			On	Fully open	Fully closed	Fully closed
Fan alarm (ON)			On	Off	Off	Fully open	Fully closed	Fully closed
Fan alarm (OFF)			Off	On	Off	Normal	Normal	Normal

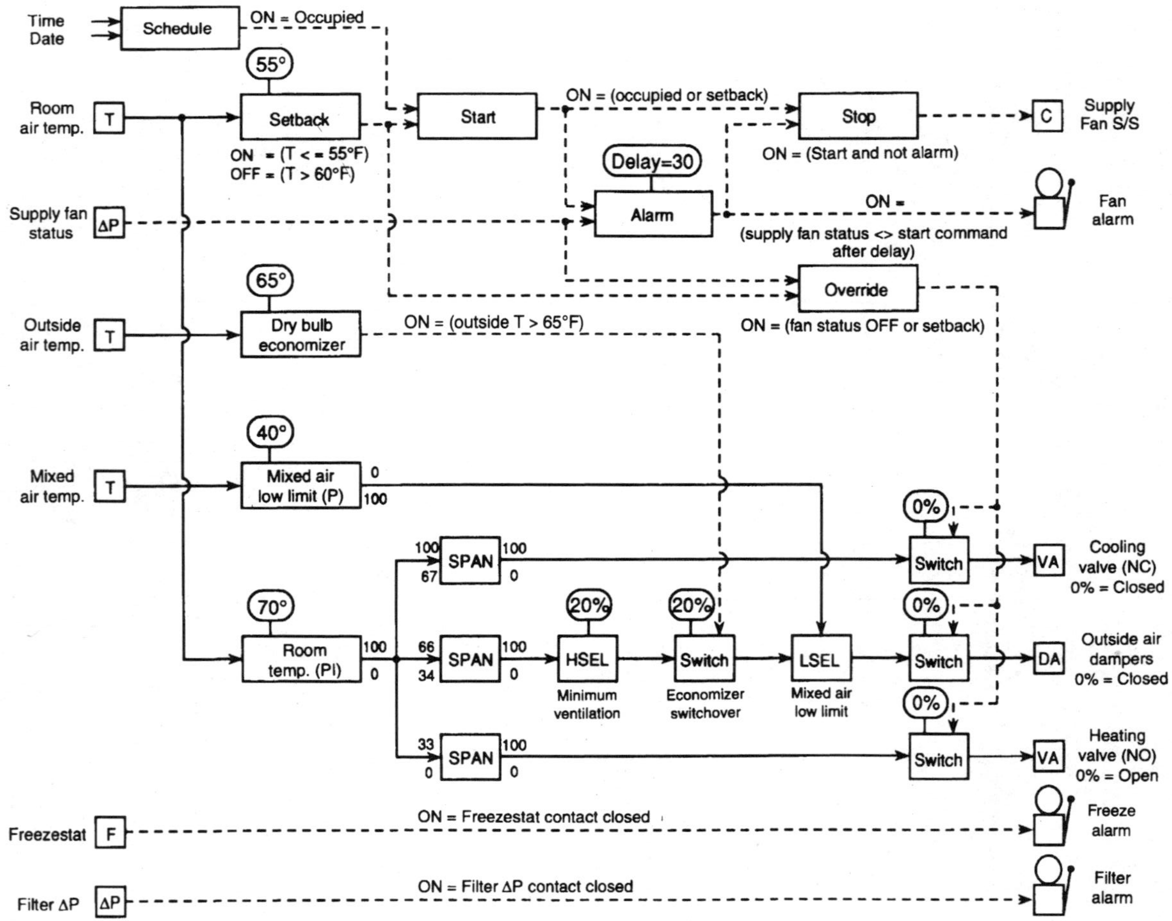

FIGURE 8.3. The completed flow diagram depicting a schematic representation of each sensor, actuator, and logic sequence.

If you study Fig. 8.3 for awhile, you will soon get comfortable with the idea of such diagrams and should see, at least generally, how the technique can be used to depict arbitrarily complex logic.

8.6 CONTROL SOFTWARE

The final step in the process is to develop the actual control software that will execute in the DDC. The configuration method will depend on the equipment to be used as has been described in Chapter 4. Now is the time to go back to the system diagram and add the AI, BI, AO, and BO blocks that show the actual connection to the DDC panel. See Fig. 8.2C. In a real job, these blocks would indicate the wiring termination to be used, e.g., DDC Panel 1, AI-2, and possibly the symbolic variable name to be used in the software to refer to the input or output such as AHU1.MAT.

If the system to be programmed happens to use graphical programming, the flow diagram can essentially be used "as is," assuming you have had the foresight to use the same logic symbols as employed by the DDC vendor's graphical programming language. If the DDC is to be programmed using a text-based language, having the flow diagram, along with the verbal sequence of operation makes the coding process relatively easy, if not trivial.

8.7 CONCLUSION

In addition to the documentation we have developed in this chapter, a designer would also have to produce an overall materials list; a wiring list showing each wire, properly labeled, along with its origination point in the field and its termination at the controller; the physical layout of the equipment, including plan views and riser diagrams; and electrical schematic diagrams of the entire installation.

While not all aspects of the design methodology presented in this chapter are industry standards, the method still provides a straightforward, disciplined way of getting on paper all of the relevant aspects of the control system's components and operational logic. As the use of computer-aided design and drafting systems becomes even more widespread, it will be possible to "recycle" flow diagrams, or portions thereof, just as it has long been possible to reuse sections of DDC control programs of proven worth. The end result will be a speedier design process and far superior documentation.

REFERENCES

ISA: S-5.1, *Instrumentation Symbols and Identification*, available from ISA, 67 Alexander Drive, Research Triangle Park, NC 27709.

SAMA: PMC 22.1, *Functional Diagramming of Instrument and Control Systems*, available from SAMA, 1101 16th Street, NW, Suite 300, Washington, DC 20036.

Chapter 9

DDC Implementation

SUMMARY

This chapter will look at some of the practical considerations involved in successfully implementing distributed DDC starting with design. Then we will look at specification writing as it applies to DDC and will consider some of the aspects of installation, commissioning, and training that are vital to success. Lastly, we will take a brief look at some of the issues pertaining to operation and maintenance.

9.1 DDC SYSTEM DESIGN: A FEW MORE THOUGHTS

As described in the previous chapter, the initial stages of designing a DDC installation are identical to those of a conventional pneumatic or analog electronic control system: the first step is to thoroughly investigate the system that is to be controlled so as to gain a complete understanding of its intended operation. In particular, don't forget to take into account the control actions to be taken as a result of malfunctions. Most malfunctions should be treated as alarm conditions and their description should include alarm thresholds, deadbands, and, possibly, even the alarm message texts that should be conveyed to an operator. In addition, you need to find out the normal setpoints to be associated with each control loop, any applicable temperature reset schedules, equipment operating schedules, any special energy management functions that need to be incorporated into the final design, and any aspects of the system that are to be monitored beyond those required for actual control. In this last category would be items like air filter alarms and metering requirements for steam, chilled water, and electricity.

In the case of a retrofit situation, it is imperative to investigate the current condition and operation of the system. There may be a need for certain repairs, even *before* the installation of new controls.

Once the desired operation of the system is understood, the next step is to figure out what sensors and actuators will be needed. Remember that point sheets are not only important in helping to determine the number and type of sensors and actuators needed, but also serve as an important asset to the person who will ultimately have to program the controller. They likewise serve to give prospective bidders a better grasp of the project scope for those jobs that are to be competitively bid.

Another important consideration is the provision of some spare capacity for future use. To be most useful, there should be spares of each input and output type, perhaps amounting to as much as 20 or 25%, in each controller or, at least, in each mechanical room. There is, of course, a natural, albeit perverse, law of the universe at work here. If you provide plenty of spares, you will probably never use them. If you don't provide any, people you've never even heard of will start clamoring for additional points!

Some thought also needs to be given to the possibility of controller failure. It might be physically possible, for example, to cram all the I/O and control software into a single box. Yet a better approach might be to divide the functions up between several controllers so that the overall system can limp along, even if one of the controllers fails. This presumes that the control strategy, and the controllers themselves, can operate satisfactorily in a distributed fashion, taking into account such factors as system time constants and intercontroller communication speeds.

9.2 SPECIFYING DDC

Writing specifications is a profession unto itself. If you want to develop some appreciation for this, along with the history and lore of spec writing, I would urge you to contact The Construction Specifications Institute (CSI) and get a copy of its *Manual of Practice*. It is not only the "bible" of spec writers, it begins by pointing out that the first specification ever recorded appears in Genesis 6:14 wherein the Lord describes to Noah the construction specifications for an ark of such materials and dimensions as to be able to withstand a most prodigious flood! Specifications have evolved in the years since as one of the principal forms of construction communications, the purpose of which is to convey the intentions of the owners and designers to those who will be doing the actual construction.

Let's look at some of the general aspects of specification development. "Specify," according to the CSI, means "to describe or define in detail." Specifications define the "qualitative requirements for products, materials, and workmanship upon which a construction contract is based." Because of this, specifications have a legal, as well as technical, nature.

Specs are but one component of the complete set of construction documents. In addition, a complete document set typically contains bidding requirements, contract forms, contract conditions, drawings, addenda, and, possibly, contract modifications. Prior to 1963, there were no standards at all for the layout and format of specifications. It was in that year that the CSI developed the 16-division format which organized construction specifications into generically related groupings. This 16-division format continues in effect to this day and is fully described in the CSI document called *MASTERFORMAT—Master List of Titles and Numbers for the Construction Industry*. See Table 9.1.

Divisions are further divided into "sections," each of which describes a particular material or product and its installation. Each section is comprised of three "parts": General, Products, and Execution. Each part is divided into numbered articles made up of one or more unnumbered paragraphs.

In an effort to achieve as high a level of standardization as possible while still allowing for the flexibility to accommodate unforeseeable specifying requirements, sections are classified as broad scope, medium scope, and narrow scope. Broad scope sections, as the name suggests, subdivide a division into broad areas of related topics which may then be further subdivided into medium and narrow scope topics. Each broad scope section is assigned a 5-digit number, the first two digits of which correspond to the division number. The numbers have been selected to allow plenty of room for adding medium and narrow scope sections beyond those suggested. MASTERFORMAT also provides a brief explanation for the type of construction activity intended to be covered by each broad scope title.

As an example, let's consider Division 15: Mechanical, the division that is intended to address mechanical systems and equipment not covered elsewhere. At present, the division is subdivided into 12 broadscope sections. See Table 9.2.

The controls section, 15950, is specifically intended to cover the equipment supplied to control the HVAC products specified in division 15 and is certainly the logical place for DDC. A note accompanying the broad scope explanation of section 15950 points out, however, that there are two other sections where controls could be specified. Section 16900 in the electrical division covers controls used solely for electrical systems (such as lighting) and section 13800 in the special construction division applies to "building automation systems," defined as controls "integrating the entire building."

TABLE 9.1 CSI Specification Divisions

Division 1: General Requirements
Division 2: Site Work
Division 3: Concrete
Division 4: Masonry
Division 5: Metals
Division 6: Wood and Plastics
Division 7: Thermal and Moisture Protection
Division 8: Doors and Windows
Division 9: Finishes
Division 10: Specialties
Division 11: Equipment
Division 12: Furnishings
Division 13: Special Construction
Division 14: Conveying Systems
Division 15: Mechanical
Division 16: Electrical

TABLE 9.2 Division 15 Broadscope Sections

15050 Basic Mechanical Materials and Methods
15250 Mechanical Insulation
15300 Fire Protection
15400 Plumbing
15500 Heating, Ventilating, and Air Conditioning
15550 Heat Generation
15650 Refrigeration
15750 Heat Transfer
15850 Air Handling
15880 Air Distribution
15950 Controls
15990 Testing, Adjusting, and Balancing

One ramification of specifying DDC in division 15 is that the controls are usually handled as a subcontract to the mechanical contractor. The building owner then finds himself dealing with the controls company through a middleman, a situation which often leads to a loss of control over the work of the subcontractor. For this reason, it is sometimes preferable to bid the controls portion of a job separately from the other mechanical work. In such cases it might be cleaner to put the controls specs in division 13 or to make a new division 17 covering controls alone.

Once you figure out where you want to put the DDC spec, the next step is to flesh out the details of the specifications. Four different methods of specifying are commonly used. The first is called the descriptive method. In this method you define the exact materials and installation techniques to be used but without the use of brands or trade names. What you are actually doing is describing the means to a desired end. An example would be specifying the use of thin-film 100-Ω platinum RTDs for the measurement of air temperature.

The second specification method is called the performance method. In this case you specify the results you are looking for and criteria for determining if the results have been met. The contractor is then free to determine what materials to use to comply with the performance criteria. Where descriptive specs specify the means to an end, performance specs specify the end itself. To continue the example started above, you could specify that the contractor shall use temperature sensors that have a resistance of 100 Ω at 0°C and a temperature coefficient of resistance of 0.00385 Ω /Ω /°C. Such a specification would lead the contractor to select a platinum RTD, but he might choose a wire-wound model rather than the thin-film style.

The third way to specify something is to reference a standard. You could say, for example, that all temperature sensors shall comply with DIN standard 43760. Since this standard defines all the characteristics of 100-Ω platinum RTDs, you would again get what you had in mind.

The fourth specification method is the proprietary method. You simply state the actual make and model of equipment to be used.

In practice, all four methods can be used in the same specification. Be aware, however, that only one method should be used for any particular item. Mixing a descriptive specification with a performance specification can be particularly dangerous. If the results are not satisfactory, the contractor can simply point to the descriptive part of the spec and say "Look, I used precisely the materials and methods you called for. I can't help it if the results are not what you wanted!"

Let's look now at how the DDC specification section can be organized. As previously pointed out, each specification section normally contains three parts, General, Products, and Execution. The General part contains administrative and procedural requirements that are unique to the section. Among these should be a summary of what the job entails, details of required submittals, and warranty requirements. The Products part of the section details your requirements for hardware, software, and communications. The Execution part lays out such things as installation practices, requirements for commissioning, and operator instruction. I've also seen instances where this part contained "generic" sequences of operation for different types of HVAC systems.

In an effort to summarize the most prominent items that a DDC spec should contain, I offer Tables 9.3 to 9.5. Each table lists the items that are usually found in

TABLE 9.3 DDC Specification Part 1: General

Summary
State overall purpose of this section
State any special design philosophy
Indicate related sections of specification, if any
References
Cite all relevant standards from ANSI, ASHRAE, EIA, IEEE, ISO
Definitions
Define all specialized or unusual terms used in the section
System Description
Indicate overall design and performance requirements
Submittals
Should include system drawings, wiring diagrams, schematic diagrams, engineering data, installation instructions, and operating instructions for each supplied component
Technical specifications for each device to be supplied including each type of sensor, signal conditioner, actuator, and field panel
Manufacturers' installation instructions for each component
Schematic of system layout, including riser diagrams, field panel layout, sensors and actuators, and interconnections
Proposed panel loading and spare capacity
Complete I/O summary
Listing of all site-specific application software, i.e., the control programs for each controller
Testing and commissioning plans
Operation and maintenance manuals
Training documentation
Contractor Qualifications
Sequencing and Scheduling
Warranty
Maintenance

TABLE 9.4 DDC Specification Part 2: Products

- Acceptable Manufacturers
- Sensors
  - For each sensor application, indicate operating range, accuracy requirements, mounting specifications, whether transmitters are desired/required
  - Temperature: Wall mounted; duct mounted; well installation.
  - Humidity: Indoor; outdoor
  - Pressure: Transducers; P/E switches
  - Flow: Air; water; steam
  - Electricity: Voltage; current; power
  - Indoor Air Quality: CO_2; particulates
- Actuators and Output Devices
  - Relays, latching or momentary, and contactors
  - E/P switches and transducers
  - Valves
  - Dampers
- Direct Digital Controllers
  - General purpose, field programmable closed loop monitoring and control
  - Application specific control: VAV, packaged AC units, zone reheat control, unit ventilators
- DDC Hardware
  - Types and quantities of inputs/outputs including pulse totalization and rate
  - Spare I/O capacity
  - Environmental requirements
    - Expected operating temperature range
    - Expected operating humidity range
    - Power conditioning requirements (possibly external)
    - Underwriters Laboratory requirements
    - Cabinet requirements relating to moisture, dust, locks
  - Failure mode operation
    - Retain last output state
    - Go to known state
    - Application program in EEPROM for automatic restart
  - Real-time clock functions
  - Battery backup
  - Local OMI features including local status of I/O with Hand/Off/Auto switch or override potentiometer
- DDC Software
  - Configuration methods: text-based, graphical
  - Required logical functions: AND,OR,NOT,XOR
  - Required math functions: SQRT,AVG, EXP, MIN, MAX
  - Software backup
  - Time-based start/stop control
  - Minimum energy start/stop optimization
  - Closed loop control, both positional and incremental PID and two position
  - Duty cycling
  - Load shedding
  - Temperature setback
  - Dry bulb or enthalpy economizer control
  - Discriminator control, supply air or water reset
  - Event-initiated programs
  - Multiple boiler or chiller part load dispatching
  - Utility metering

TABLE 9.4 (*cont.*)

DDC Software (*cont.*)
- Alarm and event annunciation/logging
- Monitoring of simultaneous heating and cooling
- Trend/history logging: rate, period or number of samples, for all points or only a limited number
- Equipment runtime logging
- Lighting control
- Fire, smoke, security functions

Communications and Networking Requirements
- Ability to share data without central computer (peer-to-peer)
- Maximum allowable alarm annunciation time
- Protocol requirements: open, standard, or proprietary
- LAN requirements: distance, number of nodes
- Cable requirements
- Remote access requirements: dial-in modem

Operator Workstation
- Input devices: keyboard, mouse, trackball, joystick, touch screen, light pen
- Color, dynamic refresh graphics
- Graphics composer program
- Archival storage requirements
- *X-Y* plotting capabilities
- Alarm display, history
- Operator security features: user id, password, privilege levels
- Printer requirements
- Database management tools, spreadsheet interface
- Field panel configuration tools: language compilers, graphical programming software, backup retention of configuration
- Upload/download software
- Interactive test programs, direct field I/O from workstation

the corresponding part of the section and contains some keywords to remind you of some of the things to consider as you draft the particular language that will have the best chance of getting you what you want!

If you want to get a better idea of some the most common language and phraseology used in DDC specs, consider contacting your favorite manufacturers, almost all of whom will be happy to provide sample specifications. Bear in mind, however, that these samples are inevitably proprietary specs that have been disguised as descriptive or performance specs for the not-so-clandestine purpose of eliminating competition! Nonetheless, they usually serve to point out what a particular vendor thinks are the distinguishing features of his equipment and, if used intelligently, can often provide some good ideas for your own applications.

Another option would be to get a copy of the NIST public domain *Guide Specification for Direct Digital Control Based Building Automation System* that was published in June 1991 for the federal government's General Services Administration.

.E 9.5 DDC Specification Part 3: Execution

,ıstallation practices
Wiring: shielded, twisted pairs, coax, conduit
EMI/RFI shielding and grounding
Cabinet penetration only from sides and bottom
Separation from power-carrying circuits
Mounting in dry, vibration-free location
Local code requirements
Inspection and testing
Startup and acceptance
Commissioning requirements
Operator training
Maintenance technician training
Programmer/configurer training
As-built documentation and drawings

9.3 INSTALLING DDC

The installation of DDC sensors, actuators, field panels, and interconnecting wiring mostly revolves around common sense and adherence to the fundamental requirements of relevant building codes. In this section I will point out a few areas where I have had personal learning experiences.

First, let's look at sensors and actuators. The biggest problem with air temperature sensors is their placement in the ductwork. The amount of stratification, i.e., hot air at the top of the duct, cold air at the bottom, can sometimes approach 10°! This depends, of course, on the type of air handler, the season of the year, the proximity of heating or cooling coils, and what is happening in the upstream ductwork in terms of mixing and turbulence. There are two methods of solving the problem. The first is to determine the best sensor location by sampling the temperature at several places in the duct by drilling small holes in the sheet metal and measuring the temperature at each position. The holes can be easily sealed and a representative position can usually be found. The second approach is to use an averaging probe. These can be 10 or 15 ft long and can be rigged across the duct in a serpentine manner to again produce a representative measurement. In the case of temperature sensors mounted in thermowells, it is usually desirable to fill the well with heat conducting grease to reduce the response time of the sensor by effectively bringing it into direct thermal contact with the fluid whose temperature is being measured.

When measuring the flow of either air or water, the sensors should be located downstream as far as possible from elbows, bends, valves, dampers, or other obstructions. If at least 10 pipe or duct diameters are not available, it may be possible to obtain believable, if not accurate, results using flow straightening baffles.

Pressure or flow switches should be mounted on nonvibrating structures if at all possible. Airflow switches mounted on the side of ducts have a tendency to jump around so much that contact bounce is a distinct possibility. This can result in intermittent status alarms as the DDC panel alternately detects made and broken contacts, even though the air handler is running normally.

DDC panels themselves should be mounted in the best environment that you can find for them. Two criteria are critical: the panel needs to be *accessible* and there should be adequate *light* so that a technician can see what he or she is doing. I once encountered a panel mounted on a wall behind a chilled water pipe in such a way that it was impossible to open the enclosure door more than about 2 inches! Remember also that digital electronics abhor water. For this reason, any conduit penetrations into the enclosure should be made through the side or, better still, through the bottom. Penetrations through the top should be avoided at all cost. I've seen several panels that were blown up because the top-entering conduits allowed condensation to drip onto the printed circuit boards. Even though the quantity of water was minute, each drop contained enough dissolved salts and minerals that eventually a conductive path was deposited on the PC board resulting in a short circuit, smoke, and flame.

Wiring needs to conform to the National Fire Protection Association's National Electric Code, for starters, along with any local codes that may be in effect. Such codes may determine if and where conduit is required, whether exposed cables have to have high-temperature jackets, and so on. If conduit is to be used, always pull several additional multiconductor cables or, at a bare minimum, leave a pull string in the conduit. The logic is the same as that which dictates leaving some I/O capacity in each DDC panel. Second, do not allow splices to be made other than at known locations, preferably in junction boxes. If you do, the splice will invariably fail and you'll have a devil of a time trying to do something about it. Of course, if you took my previous advice you will at least be able to replace the affected conductor(s).

I like to view the use of shielded wire for sensor and actuator hook up as insurance: you may be able to get away without it, but if you use unshielded wire and pick up noise signals through inductive or capacitive coupling, you won't have much choice but to replace or relocate the cable. In such cases, the use of shielded cable from the start would have to be seen as a cheap insurance policy indeed. If you do use shielded cable, make sure that the shields are only grounded at one end. Otherwise, differences in ground potential can cause currents to flow through the shield, possibly resulting in a worse situation than if unshielded cable had been used.

Electrical surge and lightning protection is another major area of concern, particularly if you are installing a system in a part of the world where thunderstorms commonly occur. The transient voltage spikes produced by lightning can easily destroy unprotected, unhardened digital circuits. On the positive side, one of the few benefits attributable to the fear of nuclear annihilation has been a steady improvement in the capability of military digital electronics to withstand *electromagnetic pulse* (EMP) disturbances of the kind that would occur if a nuclear explosion occurred somewhere within your line of sight. These same technological improvements are gradually finding their way into the less costly realm of civilian, commercial electronics where the need is protection from the EMPs produced by lightning.

There are basically three ways that transients can enter a DDC panel: on the power lines; over sensor, actuator, or communications wiring; or through the air. The best protection, therefore, involves interdicting all three potential sources. Although most second-generation DDC devices are more rugged by design than older vintage equipment, it may still be desirable to put a surge filter on the incoming AC power line. These are inexpensive and they work. You might also want to get a copy of IEEE Standard 587-80 *Guide for Surge Voltage in Low Voltage AC Power Circuits* which addresses this topic. Other circuits, particularly those that are not run in conduit or underground, might benefit from optoisolators, gas discharge protectors, or the use of metal oxide varistors (MOVs) where they interface to the DDC input circuitry. Unlike optoisolators which might or might not be provided by the original equipment manufacturer, gas discharge tubes and MOVs are almost always field installed.

9.4 COMMISSIONING DDC

Commissioning definitely means different things to different people. Traditionally, the commissioning process has been synonomous with "final checkout and acceptance testing." Depending on the procedures used, this still may be a reasonable approach. It is certainly far better than just taking the contractor's word that everything is wired and functioning flawlessly. I don't mean to suggest that contractors go around intentionally trying to deceive people; rather, it is simply a fact that a job with no crossed or mislabeled wires, no improperly located or malfunctioning sensors or actuators, and no software logic errors is virtually unheard of!

It is because of the near certainty that such problems will occur that commissioning is such a vital element in the successful installation and use of DDC. Without commissioning, for example, it is not even obvious when the contractor has successfully completed his work. Unlike most of the other construction trades, the physical installation of the components does not mean that the job has been successfully completed, i.e., that a working control system has been provided. At least half, if not three-quarters, of the job is tied up in the control logic and the software that attempts to implement it. If a thorough, comprehensive commissioning effort is not made while all the players are still on site, the nearly inevitable result will be that the system gets "commissioned" over its entire lifetime as the maintenance staff is called upon to deal with one anomaly after the other.

Given that commissioning takes time and therefore costs money, the precise contents of commissioning programs has been a matter of considerable debate. The issue was tackled head-on by an ASHRAE standards committee that developed, in 1989, the society's first "guideline" (a less "definitive" document than a standard applied to an area where there may be a variety of approaches), ASHRAE Guideline 1-1989, *Guideline for Commissioning of HVAC Systems*. We will refer to the document as G-1. G-1 has created quite a stir. Unlike the traditional concept that commissioning is something to be left until the very end of a job, G-1 envisions commissioning as an ongoing process that begins when a job is first conceived and extends until occupancy and beyond having as its purpose "a fully functional, fine-tuned HVAC system." Who could argue with that? While some consider this approach to be overkill, G-1 does set forth many useful ideas and procedures, including those normally associated with more traditional views of the process.

To begin with, G-1 introduces the concept of a *commissioning authority*. This entity, which may be the person who originally prepared the contract documents for the job, the contractor, the owner, or a committee of two or more of these, is responsible for laying out and executing the commissioning plan from beginning to end. G-1 views the commissioning plan as a document which outlines the organization, scheduling, allocation of resources, and documentation pertaining to the overall commissioning process.

G-1 then goes on to divide the process into five phases: pre-design, design, construction, acceptance, and post-acceptance. Let's look briefly at what is supposed to happen during each phase.

- *Pre-Design:* During this phase the groundwork for the entire commissioning process is established, responsibilities for specific commissioning steps are assigned, and the documentation to be used to perform benchmark testing is developed. Part of this documentation is the baseline information about the project: occupancy requirements, design assumptions, building loads/zones, and utilization.
- *Design:* During this phase documentation is prepared to describe all aspects of the HVAC systems and controls in such a way that subsequent verification procedures can be developed. This documentation includes a control system description, sequences of operation, seasonal change-over procedures, part-load strategies, occupied/unoccupied operation, and design setpoints and ranges.
- *Construction:* During construction, the HVAC system is installed, tested, and put into operation in preparation for *functional performance testing.* This is really the heart of the process. G-1 defines functional performance testing as "that full range of checks and tests carried out to determine if all components, subsystems, systems, and interfaces between systems function in accordance with the contract documents. In this context, 'function' includes all modes and sequences of control operation, all interlocks and conditional control responses, and all specified responses to abnormal emergency conditions."

G-1 emphasizes the importance of getting the operations and maintenance people involved during this phase by having them on-site, from time to time, to observe the installation process. This is something that just seems like common sense. Often, however, the first time O&M personnel get invited to look at the systems they will ultimately be responsible for is during a brief, usually woefully inadequate, "walk-through" after construction is complete.

Controls testing should begin after installation and concurrently with the traditional "testing, adjusting, and balancing" work. This testing should include: each sensor; each actuator throughout its travel; all relays, A/D, and D/A converters; all system interlocks, interconnection, and safety devices (such as freezestats); and calibration of each input and output channel.

- *Acceptance:* This is the phase of G-1 commissioning that most nearly corresponds to the traditional view. Part of the previously developed commissioning plan should be detailed checklists of each functional performance test. The checklists, of course, and the most efficient order of testing, will vary depending on the specifics of the job. As a basic principle, the testing should progress from the individual components, e.g., pumps, valves, flow switches, etc., to discrete subsystems, such as steam, glycol, chilled water, to complete systems, such as the overall air handler itself. At the end of the process, according to G-1, "every mode of systems operation, all system equipment, components and zones, and every item in the control sequence description should have been proved operational under all normal operational modes, including part and full load, and under abnormal or emergency conditions." The results of each test should be thoroughly documented as each test is run, not only for the purpose of tracking the progress of the testing, but for use later on as a benchmark for how the system behaved when it was new.

Operator training is a critical part of this phase and should involve both on-site and off-site components.

- *Post-Acceptance:* This phase amounts to follow-up. Not only should the system documentation be modified as necessary to reflect the actual "as-built" conditions, the systems should be re-tested occasionally using the original functional performance testing checklists, and documented results, as a guide. Is this pie-in-the-sky? Perhaps, but it might be a way to actually save money if the tests bring to light problems that might otherwise have gone unnoticed.

Regardless of whether or not you want to implement the entire scope of G-1, here are some essentials. Make sure someone is assigned commissioning responsibility. This person needs to develop a definitive plan, the more detailed the better, and the requirements need to be referred to in the specifications. This is where real creativity is required since some degree of simulation is almost always needed to force the controls through all their operational modes. You may have to manipulate the controller's clock, for example, to see what happens on leap year or at the transition from unoccupied to occupied schedule times. You may have to simulate temperature or other sensor extremes by temporarily replacing the sensor with a voltage or current generator or a resistance decade box, and so on. The contractors need to know that they will be expected to have people available to perform the tests and you should plan to get your O&M people involved as early as possible.

The more of these steps you are able to take, the greater will be the likelihood that the overall system will perform satisfactorily over the long haul!

9.5 TRAINING

The more your operations and maintenance people know about DDC in general, and your specific equipment in particular, the better off everyone in your organization will be.

Assuming the trainees have been identified, who should actually do the training? One option is to send the student(s) off-site to a program given by the DDC manufacturer. All of the major manufacturers, and most of the smaller ones, have such programs. The larger manufacturers have extensive facilities that rival those of a college or university and have a full-time staff of professional instructors. If you have enough people that need the training, it may be more cost-effective to bring the manufacturer's instructors on-site, although you need to be sure your people can attend without too much distraction from their normal duties.

Another option is to do the training yourself or with your own instructors. We have sometimes successfully combined these two options by sending someone to school who then comes back and passes on his or her new knowledge to colleagues.

Yet another possibility is to send the students to a short course or professional development seminar held by the various trade and professional organizations. These tend to be oriented toward general principles, however, and will need to be combined with site- and manufacturer-specific training to be of most use.

One of the major problems is the ongoing nature of the training task, a problem that is exacerbated when new operators come along. An idea that might be worth exploring is the use of a camcorder to record training sessions or to develop a suite of in-house tapes that lead a newcomer through the essentials of your particular installation. Other training aids include books like this, the ASHRAE handbook series, and self-study courses on such topics as control theory from the Instrument Society of America.

What should the training cirriculum encompass? This depends, of course, on the backgrounds of the trainees and whether they will be primarily involved with operation or maintenance. Here are some items to consider:

Fundamentals that include basic control theory, sensors and actuators, computer hardware and software, and data communications.

Site-specific information that includes general controller architecture and layout; HVAC functions performed; DDC configuration techniques; OMI functions including making changes to setpoints, overriding outputs, troubleshooting alarms, database entry and query, using reports and logs, and diagnostics.

Maintenance personnel should also be given additional specifics: Physical location of each control device, field troubleshooting and diagnostic procedures including communications, repair instructions where applicable, preventive maintenance procedures and schedules, and routine calibration procedures.

In summary, training is an essential investment in the future. There is probably no way to overdo it, and it proves to the recipients that they are viewed as important and worthwhile members of the team!

9.6 OPERATION AND MAINTENANCE

Depending on the extent of a DDC installation, *operation* may amount to nothing more than occasionally checking to see if a field panel is still powered up. At the other extreme is the operation of a full-blown facilities "nerve center" where data is constantly streaming in from the field, alarms and events are analyzed, and building energy use and comfort is proactively optimized. I would like focus more on the latter model.

The key to operational success of DDC is undoubtedly the people who are chosen to run the system. While background and training are significant factors, the single most important attribute of a successful operator is, in my opinion, a thoroughly positive attitude toward the technology and a knowledge of the systems being controlled: the person should really enjoy working with computer hardware, software, and HVAC.

The importance of attitude and HVAC experience was driven home to me some years ago as a result of a visit to a sister university. The institution in question had at the time a system that most would have charitably characterized as hideous: there were no color graphics, no tabular reports or trend logs, data had to be entered and interpreted in hexadecimal format, and the central computer, which had been provided by an aerospace firm specializing in rocket guidance systems, was actually attempting to perform centralized DDC! Nonetheless, the university was deriving tremendous benefit from the system because the lead operator happened to love his job. It turned out that he had once been the foreman of the temperature controls shop, had been somewhat disabled from a fall off a ladder, and was given the choice of early retirement or running the EMCS. He chose the latter and, once he learned how to deal with all the user-unfriendliness, the computer essentially became an extension of his own psyche, untiring and able to be in more than one place at once. When an alarm came in, he was able to direct a field mechanic to the exact spot of the problem, complete with advice borne of his own lengthy field experience such as "watch out you don't bump your head on the steam trap when you climb up to check out that temperature gauge!"

The moral of the story is clear. Get the right people on the job and give them the tools appropriate to their task. But if the choice is between the tools and the people, get the people first!

Many facilities owners choose to rely on the controls suppliers to some degree for operation, often through dialup modems, and to an even greater degree for maintenance. It is easy, albeit often quite expensive, to simply extend the usual one-year warranty period with an appropriate service contract. Whether this makes sense for you depends on the extent and criticality of your installation, the time delay that can be tolerated after a problem is detected and a mechanic arrives on site, and the availability and qualifications of your own people.

If you choose to do your own maintenance you will need to stock your own spare parts, including DDC processors, communications boards, power supplies, etc., and have available the proper complement of tools. A major difference between maintaining conventional pneumatic controls and DDC is the set of tools and test equipment with which the field mechanic must deal.

The most fundamental diagnostic tool is undoubtedly the battery-powered, handheld digital voltmeter, or DVM. If possible, get one with at least four and a half digits of display resolution so that you can directly read milliamps or millivolts. (A "half digit" is a digit that either displays a 0 or a 1. A $4\frac{1}{2}$digit DVM set to the 10-mA scale, for example, reads from 00.000 to 19.999.) Most DVMs can also measure resistance and can therefore be used to check continuity.

For checking sensor input calibration, it is useful to have a precision voltage or current generator. Such calibrators may also be battery-powered for easy use in the field. For verifying the calibration of RTD transmitters or input channels, a precision resistance decade box should be acquired. These have selector knobs for each digit in the output resistance value and can have a resolution down to 0.01 Ω.

Another extremely powerful piece of test equipment is the portable oscilloscope. An oscilloscope is simply a CRT with a calibrated scale that allows you to *see* voltage versus time. Moreover, the display can either run continuously or be triggered by the input voltage crossing an adjustable threshold. Analog storage scopes, or the new digital scopes, allow you to capture infrequent events such as low-frequency pulses. Besides being able to literally see the presence of electrical noise on a signal line, you can also make AC and DC voltage measurements, although not to the accuracy of a DVM.

If you will be getting involved with troubleshooting communication lines, the oscilloscope will also be a principal ally since it lets you instantaneously determine the presence or absence, and amplitude of a communication signal along with noise. If you are dealing with voice frequency modems, i.e., modems using tones in the 300- to 3000-Hz range, an audio test set will also come in handy. These devices usually have an audio amplifier and circuitry to measure both the frequency and amplitude of audio tones. As a result you can both hear and quantify the communication signal.

If you have to deal with EIA-232 connections, you may want to get an appropriate crimping tool so that you can quickly make up custom male and female connectors. Kits are available with the various pins, sockets, and shells. An EIA-232 *breakout* box can also come in handy. Breakout boxes have LEDs that tell you the state of each of the major interchange circuits and let you temporarily interconnect two EIA-232 devices so that you can verify exactly which control and data signals are needed to make the two devices work together. You can also use the breakout box to let you gain temporary access to these signals for display or testing.

If you ever get involved in troubleshooting communication software, a data analyzer or protocol analyzer is essential. Data analyzers are machines that can be bridged across a communication circuit and let you see each character that has been sent or received. This can be extremely useful for observing "unprintable" characters like ASCII carriage return, line feed, or other control characters. If you have to troubleshoot a high-speed LAN, a protocol analyzer can interpret, as well as display, the communications traffic.

To summarize, DDC maintenance requires basic electronics skills and knowledge of a different sort of test equipment than previously used with pneumatics. Once the tools and test equipment procedures are mastered and the technician understands the operation of the specific equipment at hand, problems with DDC systems can usually be solved in fairly short order.

REFERENCES

ASHRAE Guideline 1-1989, *Guideline for Commissioning of HVAC Systems*, Atlanta: ASHRAE, 1989.

CSI, *CSI Manual of Practice*, Alexandria, VA: The Construction Specifications Institute, 1988.

IEEE Standard 587-80, *Guide for Surge Voltage in Low Voltage AC Power Circuits*, New York: IEEE, 1980.

NISTIR 4606, *Guide Specification for Direct Digital Control Based Building Automation System*, 1991. Available from the National Technical Information Service, Springfield, VA 22161.

Epilogue: The Future of DDC

The future of direct digital control is bright for one main reason: the future of the underlying, enabling technologies is brighter still. In these final few paragraphs I would like to suggest some of things we are likely to see in the decade leading up to the year 2000.

Computer technology, particularly semiconductor microminiaturization, will continue to progress. Microprocessors will become smaller, faster, and cheaper. Along with this, computer memory will continue to evolve so that more and more bits will reside in increasingly smaller packages. This means that ever more sophisticated programs will be able to be stored and run.

An example of what this will mean will be the universal availability of self-tuning or adaptive controllers. The controllers of the future will be able to continuously adjust their computational algorithms to accommodate both short- and long-term changes in heating or cooling load or supply conditions. While there are several machines on the market today that implement adaptive control, the computational requirements are so great that such capabilities are currently limited to only the most expensive and sophisticated controllers.

We can also expect future DDCs to make increasing use of fuzzy logic concepts. Most control algorithms today are based on strictly Boolean logic: IF (Expression is TRUE) THEN (Do This). The Boolean "Expression" is a member of a set of expressions which are either TRUE or FALSE. There is no middle ground. In fuzzy logic, which is really a misnomer since it is firmly grounded in mathematical theory, an expression may have varying degrees of membership in a set. For example, a temperature may at the same time be 80% "warm" and 20% "somewhat cool." By scaling the controller's response, perhaps by smoothly varying the tuning parameters between those appropriate for 100% "warm" and 100% "somewhat cool," it may be possible to achieve better control across the entire temperature range. Fuzzy logic is already being used in machines as diverse as autofocusing cameras to vacuum cleaners!

Data communication standards will be globally accepted and applied. Whether this turns out, in the case of building system DDC, to be BACnet, BACnet II, or something entirely different is unimportant. The main thing is that we can expect nearly universal connectivity. It will, at last, be possible to gather information from any vendor's equipment and view it on a common display.

DDCs will become increasingly fault-tolerant and robust. Controllers today are already starting to look more and more like their industrial process control counterparts. They will be less susceptible to electromagnetic and radio frequency interference, for example, and may have redundant memory modules or even redundant microprocessors.

Configuration and programming will continue to become easier and there may even be a push for some type of standard programming methodology. Ironically, the graphical programming tools that currently distinguish several of the manufacturers from their peers, may furnish just such a basis for standardization. By providing a pictorial representation of the control logic, graphical programming effectively isolates the programmer from the underlying operation of the controller. If standard graphical symbols for all the basic control functions could be agreed to, it would be fairly straightforward to map or translate the graphical depiction into some form intelligible to the underlying machine. This is probably more of a political than technical challenge, though no less formidable as a result!

I would also not be surprised to see some type of documentation standard for indicating what control strategy is embedded in any particular control program. This could happen even in the absence of configuration and programming standards although such standards would greatly facilitate, and be a part of, improved documentation.

In conclusion, don't get discouraged if you don't understand everything there is to know about DDC all at once. No one knows it all. Just take things one step at a time, keep learning, and have fun!

Glossary of DDC Abbreviations, Acronyms, and Terms

Alarm A condition, often derived from a sensor input or inputs, outside the range defined as "normal."

Algorithm A step-by-step procedure for solving a problem, often by repetitive operations.

Analog Describes a physical quantity, e.g., temperature, that varies through a continuous range of values.

A/D Converter A circuit or device whose input is information in analog form and whose output is the same information expressed in digital, i.e., numerical form.

ALU Arithmetic logic unit. The circuitry in a computer that performs arithmetic and Boolean logic calculations.

ANSI American National Standards Institute.

Application Service A communication service provided to an application by the top layer of the OSI protocol stack.

Application-Specific Controller A DDC tailored to a particular application such as VAV box or unit ventilator control. Usually has a limited I/O capacity and a fixed program.

Architecture The organizational structure of a system including the interrelationships of software and/or hardware.

ASC *See* Application-Specific Controller.

ARCNET Attached Resource Computer Network. A token-passing LAN technology developed by the Datapoint Corporation.

ASCII American Standard Code for Information Interchange.

ASHRAE American Society of Heating, Refrigerating and Air-Conditioning Engineers

Assembler A utility program that inputs a file of mnemonically encoded instructions for a particular processor and outputs the equivalent binary-encoded machine instructions. *See* Object Code.

Assembly Language A mnemonic language for encoding computer programs for a specific processor.

Asynchronous Communications Data transmission in which each message block or character is sent without any predetermined timing relationship to the preceding block or character.

BACnet ASHRAE's Building Automation and Control Networking protocol.

BAS Building automation system.

BASIC Beginners All-Purpose Symbolic Instruction Code. A high-level, English-like computer language.

Baud A unit of signaling speed equal to the number of discrete conditions, or signal events, occurring each second; often, but not necessarily, synonymous with bits per second.

Binary A physical quantity or signal that is represented by one of two states: on/off, 1/0, high/low, open/closed, yes/no. Also, a numbering system based on the digits 0 and 1.

B/IP PAD BACnet/Internet Protocol Packet Assembler-Disassembler. A device for encapsulating or decapsulating BACnet messages that are to be conveyed using IP. The same concept can be used for other network layer protocols such as Novell's IPX.

Bit Contraction of the words *binary digit*. The smallest unit of information which can be represented.

BNC Connector A type of coaxial cable connector featuring a quarter twist, positive locking action.

Boiler Optimization A procedure for adjusting a load, split among several boilers operating in parallel, to achieve optimum energy efficiency.

Bridge A device that connects two similar physical LAN segments at the data link layer and may also perform message filtering based on LAN addresses.

Bug A computer software programming error. *See* Debugger.

Bus A network topology in which all nodes are linearly connected to the same physical segment. Also, inside a computer, a group of parallel conductors or circuits that convey address, data, or control information.

Byte A sequence of adjacent binary digits operated upon as a unit, commonly, but not necessarily, eight in number.

C A high-level, modular programming language developed by AT&T and intended to be highly portable. C compilers exist for all major computer types and there is now an ANSI C-language standard.

CCITT The Consultative Committee on International Telegraphy and Telephony of the International Telecommunications Union based in Switzerland. A major international computer standards organization responsible for the "V." and "X." standards.

Central Processing Unit (CPU) The portion of a computer that performs the interpretation and execution of instructions. It does not include memory or input/output devices.

Chiller Optimization A procedure for adjusting a load, split among several chillers operating in parallel, to achieve optimum energy efficiency.

Compiler A computer program which translates statements written in a high-level computer language into multiple machine instructions capable of being executed by a specific computer.

Control Relay An electromagnetic device for regulating the operation of a machine or device.

Control Points Actuators through which remote control functions are accomplished.

CRT Cathode Ray Tube. *See* VDT.

CSMA Carrier Sense Multiple Access. An access method in which computers contend for the use of a network by sensing the absence of a carrier signal on the communications link before transmitting. *See* Token Passing.

CSMA/CD Carrier Sense Multiple Access with Collision Detection. The medium access control technique used by most contention networks including Ethernet.

D/A Converter A device which converts a digital value into an analog signal such as a voltage or current proportional to the digital input.

Daisy Chain A serially connected LAN where each node is directly connected to a preceding node.

DARPA Defense Advanced Research Projects Agency of the U.S. Department of Defense. The organization responsible for the development of the TCP/IP protocol suite.

Data Link The physical and logical method by which electrical signals are transmitted between computing devices in a monitoring and control system. Also known as data highway, communications trunk.

Day-Type A type of day which defines a particular facility operating pattern or schedule for energy management purposes, e.g., Weekday, Saturday, Sunday, Holiday.

DCE Data communications equipment. Includes such devices as modems and line drivers.

DCS *See* Distributed Control System. An acronym used primarily in the industrial process control industry.

DDC Direct Digital Control. Closed-loop, modulating control of an output device by means of a digital computer. Also, direct digital controller, the machine carrying out direct digital control.

Debugger A utility program that allows computer programs to be stopped or stepped through one instruction at a time so that variables may be inspected and control flow altered in order to locate programming errors, or bugs.

Default In computing, a value or device to be used or an action to be taken if no alternative is explicitly provided.

Demand Limiting A control strategy where building electrical loads are turned off on a priority or rotating basis in order to keep the total building peak electrical demand below some predefined limit.

DGP Data Gathering Panel. *See* Field Computer.

Digital Pertains to the representation of a physical quantity as a whole, integral number. An 8-bit byte, for example, can represent the numbers from 0 to 255.

DIP Dual In-Line Package. An integrated circuit where the connections to the IC are made through parallel rows of pins.

Distributed Control System A control system in which control computations are made at different locations and the results coordinated. *See* Distributed Processing.

Distributed Processing The practice of dividing computational tasks among a number of discrete computers, each with its own program. Microcomputer controllers, for example, can be programmed to collect, process, and store sensor data and perform control actions at remote locations while still communicating with each other or with a central computer.

Downloading The transfer of appropriate data from a host computer to a field computer, so that control action can be implemented by the field computer.

DTE Data Terminal Equipment. Equipment such as computers or operator terminals in which information either originates or resides.

DTM Data Transmission Medium. The physical medium over or through which data communication is carried out, e.g., twisted pairs, coax, fiber optic cable, space, etc.

Duty Cycling An energy management strategy in which loads are turned off and on to reduce electricity usage.

EBCDIC Expanded Binary Coded Decimal Interchange Code. A character set code used in IBM mainframe computers. *See* ASCII.

Economizer Cycle An energy-saving strategy that compares return and outside air temperatures to determine which source will require the least space conditioning energy.

EIA Electronic Industries Association, a prominent standards developer.

EMCS Energy Management and Control System.

EMS Energy Management System.

Energy Audit The assessment of energy usage and need, for the purpose of configuring an energy management system, designing strategies for operation, and projecting cost savings.

Enthalpy Control An optimizing function that allows use of outside air for cooling when the total heat content of the outside air is less than that of the inside air.

EPROM Erasable Programmable Read-Only Memory. A form of non-volatile semiconductor memory which can be erased and reprogrammed. There are EPROMs that can be erased electrically (EEPROM) or by means of ultraviolet light (UVPROM).

Event A change in physical state detected by a sensor.

Event-initiated An action initiated by an event rather than by time criteria.

FID Field interface device. *See* Field Computer.

Field Computer A remotely located computer that monitors and controls a specific area of a facility. The control can be direct digital or supervisory and it can function stand-alone or connected to a host computer. Also known as a data gathering panel (DGP), field interface device (FID), local control unit (LCU), field cabinet, panel, remote terminal unit (RTU), etc.

Firmware Computer instructions resident in some form of ROM. In the case of most DDC systems, the entire operating system is in firmware.

FORTRAN FORmula TRANslation language. A high-level computer language originally designed to carry out scientific calculations.

Frame A computer message or group of bytes usually delimited by a header and appended error-checking information.

Full Duplex Capable of simultaneous two-way communication.

Gateway A device that connects dissimilar networks, possibly performing protocol translations.

Half Duplex Capable of alternating two-way communication.

Hardware Physical equipment such as a CPU, memory, peripherals, sensors, actuators, etc.

Hertz A unit of frequency equal to one cycle per second, abbreviated Hz. Also kilohertz, Megahertz, etc.

Host Computer A computer in a central location that coordinates information from the various intelligent field devices. Also know as the central or head-end computer.

Hub A device used to interconnect multiple LAN segments that are arranged in a star configuration.

HVAC Heating, Ventilating, and Air Conditioning.

IC Integrated Circuit. A semiconductor device containing multiple components on a single substrate.

IEEE Institute of Electrical and Electronics Engineers. A major U.S. computer standards organization.

IFID Intelligent Field Interface Device. *See* Field Computer.

IMUX Intelligent multiplexer. *See* Field Computer.

Intermittent Hardware/Software Problems Problems that occur apparently at random and cannot be replicated by any known sequence of events. Also referred to as glitches.

I/O Input/Output. The process of getting information into or out of a computer or computer peripheral.

Interactive A method by which an operator can communicate with a computer with the machine prompting or otherwise assisting while continuing to perform its other tasks as required.

Interoperate To be able to meaningfully exchange data for the purpose of carrying out cooperative computing tasks.

Interpreter A computer language translator which converts individual source statements into machine instructions by translating and executing each statement as it is encountered in real-time.

Interrupt A signal to a computer's operating system to suspend its current processing in order to perform a more important or more time-critical task.

IP Internet Protocol. A network layer protocol used to route messages worldwide. *See* TCP/IP.

ISA Instrument Society of America. Developer of numerous process control standards.

ISO International Organization for Standardization. One of the predominant groups working to arrive at international data communications standards. *See* OSI.

Large Scale Integration (LSI) The technology of manufacturing semiconductor integrated circuits (ICs) capable of carrying out complex functions. LSI ICs may contain the equivalent of hundreds or even thousands of discrete components such as resistors, capacitors, and transistors.

Line Driver A type of data communication device that passes serial data over a limited distance and requires dc continuity to operate. Also called a limited distance modem.

Linker A utility program that combines previously assembled or compiled software object modules into an executable program.

LLC Logical Link Control. A data link protocol developed for use with the IEEE 802 series of LAN specifications.

Load An energy-consuming device or system.

Load Shedding An optimizing function that cycles electrical loads off and on to keep energy use below the electric demand limit. *See* Duty Cycling and Demand Limiting.

Local Area Network (LAN) A network of computers usually characterized by relatively high-speed data communications (> 1 Mbps) and a relatively limited distance between the most widely removed machines (several thousand feet). Another characteristic is that most LANs generally involve peer-to-peer communications, i.e., there is no master station required for the network to function.

Local Loop Control The control of HVAC devices such as valves and dampers by some predetermined algorithm, independent of whether the control is carried out digitally, electronically, or pneumatically.

LVDT Linear Variable Differential Transformer. A type of transducer used to produce a signal proportional to linear displacement. Often used in conjunction with pressure sensors.

MAC Address Medium Access Control Address. A numeric value used to uniquely identify a computer node on a LAN or other network.

Machine Language The binary codes corresponding to the instruction set of a particular CPU.

Manchester Encoding An signaling technique characterized by a voltage level transition in the center of each bit time. Used by Ethernet.

Memory Devices that can store binary information in such a manner that a single bit or group of bits can be stored, accessed, and retrieved.

Microcomputer A complete computer system based on a microprocessor including memory, I/O capability, etc.

Microcontroller A microprocessor with timers, counters, and binary and analog I/O capability on a single chip.

Microprocessor A CPU on a single semiconductor chip.

Modem MOdulator-DEModulator. A device used to communicate computer information over telephone lines or other analog media.

Multitasking A sort of time-sharing procedure by which a computer can be made to perform a number of program tasks concurrently.

MUX Multiplex unit. A field device that can scan sensors, control relays, totalized power meter data, etc. but does not possess stand-alone intelligence. A MUX takes a single input and directs it to one of many outputs or takes one of many inputs and directs it to a single output, usually under some form of program control.

Network Access Method The means by which a computing device gains the exclusive use of a data link for the purpose of transmitting data.

Network Visible Data that is accessible through the use of network communication services.

Node A computer connected to a communication network.

Non-volatile Memory Memory which retains information in the absence of applied power, e.g., ROM.

NRZ NonReturn to Zero. A method of electrically encoding 0's and 1's where the signal remains at a low or high signal level for the entire duration of the bit.

Object In BACnet, a collection of related information used to describe in a network-visible way the functioning of some aspect of a DDC.

Object Code The machine language instructions generated by an assembler or compiler.

Octet A sequence of adjacent binary digits, operated upon as a unit, that is exactly eight in number. *See* Byte.

OMI Operator-Machine Interface, e.g., CRT, printer, portable programming unit, etc., and the software to operate it. The means by which a human operator interacts with a computer system for the purpose of inputting data or obtaining information.

Open Systems Interconnection A set of data communications procedures proposed by the ISO in which the problem of communication between dissimilar machines is divided into seven discrete functional entities called layers.

Operating System A generally complex software system which manages a computer and its peripherals and allows for human interaction.

Optimizing Functions Functions which improve the efficiency of an HVAC system by automatic, real-time adjustment of system equipment.

Optimum Start/Stop An optimizing function that saves energy by starting or stopping systems after evaluating such factors as temperature, weather conditions, thermal inertia of the building, and hours of occupancy.

OSI *See* Open Systems Interconnection.

Packet A computer message containing protocol control information and ready for network conveyance.

Parameter A value supplied to a calculation. The value may be a constant or a variable depending on its use within the calculation.

PCI Protocol control information. Information necessary for the operation of a particular protocol. PCI often precedes data in a PDU.

PDU Protocol data unit. An OSI term describing a collection of octets containing PCI and, possibly, data.

PID Control Proportional, Integral, Derivative Control. Generation of a control output signal proportional to an error signal equal to the difference between a measured variable and its setpoint, the length of time the error signal has existed, and its rate of change.

Pixel Picture element. The smallest area of a CRT display that can be accessed and illuminated or colored.

PLC *See* Programmable Logic Controller. *See also* Powerline Carrier.

Plug Compatible A device that has precisely the same electrical and operational characteristics as another and can therefore be directly substituted for it without wiring or software changes. No currently available DDCs fit this description.

PMI Person-Machine Interface. *See* OMI.

Polling The data communication technique wherein a master station invites a subordinate slave station to send any data it may have queued for transmission.

Powerline Carrier A data communications technique using the AC powerlines as the physical transmission medium.

Program A sequence of instructions causing a computer to perform a specified function.

Programmable Logic Controller A digital controller originally designed to replace discrete component ladder logic controls. Primarily used to control industrial processes.

Protocol A set of conventions governing the orderly communication of data between devices on a data link. May include both hardware and software components.

Protocol Stack A set of layered protocols. *See* Open Systems Interconnection.

Pseudo code The result of subjecting computer source code to a translation process that produces a compressed version of the instructions which are then interpreted in realtime by the CPU. *See* Object Code.

PTP The Point-to-Point Protocol specified in ASHRAE's BACnet.

RAM Random-Access Memory. Volatile computer memory. Memory information is lost when power is removed from the computer.

Real-time Pertains to computations made at the time they are actually needed by a physical or other computational process.

Repeater A device that connects two physical LAN segments and exactly reproduces the electrical signals received on one segment on the other.

Ring A circular LAN configuration where messages are passed from one node to the next until they finally return to the originator.

ROM Read-Only Memory. Nonvolatile computer memory. Memory information is retained when power is removed from the computer.

Router A device that connects two or more networks at the network layer. Routers perform path selection and message relaying.

Scanning Repeated checking for a change in the physical conditions sensed by sensors.

Scheduling Programming the control of loads based on time.

Segment A single contiguous network medium—wire, fiber optic cable, etc.—to which computers are attached. Also, a portion of a lengthy message which has been subdivided for transmission due to network limitations on the maximum message length that can be conveyed.

Software A general term applying to computer programs whether written in machine, assembler, or high-level languages and without regard for the program's specific purpose; the part of a computer system that is not hardware.

Source Code A computer program written in a human-readable form such as BASIC, C, Pascal, or some form of assembly language.

Star A network configuration where segments are interconnected through a central hub.

Station *See* Node.

Strategy A philosophy of operation of a facility, or portion of a facility, to achieve the design objectives.

Strobe A process by which data on a parallel bus are read or written based on the occurrence of an enabling pulse.

Supervisory Control The control of equipment based on global considerations such as time of day or overall electrical demand. It is generally applied to entire HVAC systems as opposed to the control of individual actuators. Supervisory control was originally only possible from a centralized location; with the advent of intelligent field panels, supervisory functions can be carried out in the absence of a central computer. *See* Local Loop Control.

SVGA Super VGA. A high-resolution graphic interface used in IBM computers, typically 1024 × 768 pixels with 256 colors.

TCP/IP Transmission Control Protocol/Internet Protocol. A widely used group of protocols developed by DARPA in the United States which form the basis for the global network known as the Internet.

Token-Passing A communications network access method in which a special message, the token, is passed from one controlling device to another to establish which device has control of the data link.

T/P Token-passing.

Translator A utility program that takes a source file and generates a compressed, binary version of the file. The latter is called pseudo code.

Trend-log A record of system parameters that is updated frequently enough to establish trends in the variable being measured, e.g., temperature or energy use.

UART Universal Asynchronous Receiver Transmitter. An IC capable of carrying out serial asynchronous communication.

UDP User Datagram Protocol. One of the DARPA protocols that allows the sending of unacknowledged blocks of binary data to nodes accessible using the IP protocol.

Unitary Controller A DDC used to control a single piece of equipment such as a chiller or heat pump. *See* Application-Specific Controller.

Uploading The transfer of data, such as the current program or database, from a remote computer to a host computer.

USART Universal Synchronous Asynchronous Receiver Transmitter. Like a UART but with timer circuitry to also allow serial synchronous communication.

VDT Video Display Terminal.

VGA Video Graphics Adapter. An interface board that allows an IBM PC to display graphic information in pixel format. The typical resolution is 640 × 480 pixels with 16 colors. *See* SVGA.

VLSI Very Large Scale Integration. Semiconductor fabrication techniques that permit a very high density of components in an IC.

Volatile Memory Computer memory, e.g., RAM, in which the stored digital data are lost when power is removed.

Wide Area Networks (WAN) Computer networks characterized by relatively slow speed (< 19.2 Kbps) and relatively unlimited distances between the most widely removed machines. Many DDC networks still use WAN techniques, such as voice-grade telephone modems, even though the distances between machines are small enough that LAN techniques could be employed. As time goes on, the characteristic of low speed is disappearing.

XON, XOFF The ASCII characters DC3 (X′11′) and DC1 (X′13′) used for flow control on many computer data links such as between terminals and modems.

Manufacturers of DDC Equipment

AET
77 Accord Park
Norwell, MA 02061

Alerton Technologies, Inc.
2475 140th Avenue N.E.
Bellevue, WA 98005

American Auto-Matrix
One Technology Drive
Export, PA 15632

The Anderson Cornelius Company
6750 Shady Oak Road
Eden Prairie, MN 55344

Andover Controls Corp.
300 Brickstone Square
Andover, MA 01810

ASI Controls
2202 Camino Ramon
San Ramon, CA 94583

Automated Logic Corp.
1283 Kennestone Circle
Marietta, GA 30066

Automation Technology, Inc.
5280 W. 74th Street
Minneapolis, MN 55435

Barber-Colman Company
1354 Clifford Avenue
Loves Park, IL 61132

Barrington Systems
1160 Industrial Road
San Carlos, CA 94070

Cetek Systems, Inc.
1701 Junction Ct.
San Jose, CA 95112

Climate Master
7300 SW 44th Street
Oklahoma City, OK 73179

Control Pak Corp.
23840 Industrial Park Drive
Farmington Hill, MI 48024

Control Systems International
1625 W. Crosby Road
Carrollton, TX 75006

Echelon Corporation
4015 Miranda Avenue
Palo Alto, CA 94304

Electronic Systems USA, Inc.
1014 E. Broadway
Louisville, KY 40204

Enercon Data Corp.
7117 Ohms Lane
Minneapolis, MN 55435

Energyline Corp.
2065 Kittridge Street
Suite A
Berkeley, CA 94704

Functional Devices Inc.
310 S. Union Street
Russiaville, IN 46979

Honeywell Inc.
Building Systems Division
Honeywell Plaza West
Minneapolis, MN 55408

Integrated Energy Controls
7464 West 78th Street
Minneapolis, MN 55435

Johnson Controls, Inc.
507 East Michigan Street
Post Office Box 423
Milwaukee, WI 53201

Landis and Gyr Powers, Inc.
1000 Deerfield Parkway
Buffalo Grove, IL 60089

Novar Controls Corporation
24 Brown Road
Barberton, OH 44203

Paragon Electric Company
P.O. Box 28
Two Rivers, WI 54241

Sachs Energy Management Systems
P.O. Box 96
St. Louis, MO 63166

Scientific Atlanta
4300 N.E. Expressway
Atlanta, GA 30340

Siebe Environmental Controls
P.O. Box 27606
Richmond, VA 23261

Solidyne
1202 Carnegie Street
Rolling Meadows, IL 60008

Staefa Control System Inc.
8515 Miralani Drive
San Diego, CA 92126

Teletrol Inc.
340 Commercial Street
Manchester, NH 03101

Tour and Anderson, Inc.
P.O. Box 2519
Stamford, CT 06906

The Trane Company
Building Automation Systems Division
20 Yorkton Court
St. Paul, MN 55117

Triangle Controls Inc.
1180 Mira Loma Way, Suite K
Sunnyvale, CA 94086

TSBA Controls, Inc.
38-30 Woodside Avenue
Long Island City, NY 11104

Universal Software Corporation
1030 McConville Road, Suite 2
Lynchburg, VA 24502

Wattmaster 1401
Ott Road
Shawnee, KS 66061

York International
P.O. Box 1592
York, PA 17405

Index

Page numbers in *italics* refer to the principal discussion of a topic.
Page numbers in **bold** refer to the definition of a term.

A/D, *see* analog-to-digital
Actuators, 2, 3, 6, 7, 11, *46–51*, 54, 63, 66, 75, 78, 81, 98, 156, 195, 200, 202, 208, 214–217, 219, 227, 229, 233
Adaptive control, 91, 101, 223
Air quality, indoor (IAQ), 40–41, 51
Airflow, 11, 32, 34–35, 41, 51, 77, 81, 214
Alarm processing, 5, 79, 80–81, 92–94, 97–98, 100, 104, 109–111, 129, 144, 147, 149, 157, 159, 163–165, 192, 198–202, 207, 215, 219–220, **225**
Algorithm, **225**, 230
  BACnet event, 164
  BACnet network layer relay, 179–180
American Auto-Matrix, 85
Analog inputs, *11–46*, 66, 79, 155–156, 163, 200
Analog outputs, 37, *46–51*, 66, 162, 200, 202
Analog-to-digital (A/D), **42**
  conversion, 7, *42–44*, 48, 58, 66
  converter, 42, **225**
ANSI, 22, 67, 84, 133, 135–136, 139, 151, 170, **225**, 225–226
ANSI standards:
  ANSI/ATA 878.1 (1992), *ARCNET Local Area Network Standard*, 151
  ANSI/MC 96.1-1982, *Temperature Measurement Thermocouples*, 22, 51
  ANSI X3.4-1977, *American National Standard Code for Information Interchange*, 67
Application layer, OSI, 110–111, 148, *149–150*, 154, 158–159, 177
Application program interface (API), 149, 160
Application service, 149, 158, 170, 191, **225**
Application software, 70, *76–81*
Application-specific controller (ASC), 8, 65–66, 78, 81, 127, **225**
Application-specific integrated circuit (ASIC), 54
Architecture:
  computer, 1, 61, 69, 76, 219
  network, 91, 93, 148, 154, 169
ARCNET, 121, 130–131, *136–139*, 145, 147, 150–151, 174, 176, 182, 184, 192, **225**
Artificial intelligence (AI), 91. *See also* fuzzy logic, knowledge-based engineering
ASCII, 55, **56**, 58, 145, 148, 166, 169, 181, 187, 221, **225**

ASHRAE, 8, 41, 51, 89, 97, 105, 135, 149, 153–154, 193, 195, 202, 216, 219, 222, **225**
ASHRAE standards:
  ASHRAE Guideline 1, *Guideline for Commissioning of HVAC Systems*, *216–218*, 222
  ASHRAE Standard 62-1989, *Ventilation for Acceptable Indoor Air Quality,* 41, 51
  ASHRAE Standard 135P, *BACnet—A Data Communication Protocol for Building Automation and Control Networks*, 193. *See* BACnet
Assembler, 74, **225**
Assembly language, 67, 84, **225**
Asynchronous communication, 94, *113–114*, 125, 127, 145, 182, 186, **225**, 233
Avalanche photo detector (APD), 122

Backup, 82, *88–89*, 92, 96
BACnet, 8, 97, 135, 149–150, *153–193*, 224, **226**, 231–232
  conformance and specification, *190–193*
  encoding, *168–174*
  extensibility, *189–190*
  IP and IPX with, *181–182*
  lower layers, *174–189*
    MS/TP (master-slave/token-passing), *182–186*
    network layer, *175–182*
    PTP (point-to-point), *186–189*
  modeling DDCs, *154*
  objects, *155–158*
  services, *158–168*
BALCO, 14
Baseband signaling, 113, 120, 133, 139–140
BASIC, 84–85, **226**, 232
Baud rate, 114, **117**, 127, 154, 184
Baudot code, 58
Bell System, 3002 conditioned lines, 116. *See also* modems
Bernoulli's equation, **30**, 32–33
Binary inputs, 47, 156, 174, 202
Binary outputs, 12, 46–47, 50, 65, 156, 162, 200, 202
B/IP PAD, 181–182, **226**
BNC connector, 136, 140, **226**
Boiler optimization, 78, **226**
Bridge, network, *146*, **226**
Broadband signaling, 113, 120, 133, 140–141
Building control, history of, 2–8
Bus:
  computer, 43, 53, 60–61, 63, 113, 127, **226**, 232
  topology, LAN, 66, 131–133, 137–138, 140–143, 183–184, **226**

C programming language, 72, *84*, 160, 168–169, **226**, 232
Cable, *117–124*
  coaxial, *120–121*
  fiber optic, *121–124*
  twisted pair, *119–120*
CAL1, 85–86
Canadian Standards Association (CSA) Committee T530, 81
Carpenter Technology Corporation, 14
Carrier sense multiple access (CSMA) with collision detection (CSMA/CD) 129–130, 133, 135, 139, 141, 143–144, 146, 151, **227**
CCITT, 117, **125**, 185, 188, **226**
Central processing unit, 59, 61, 80, 86, **226**
Chilled mirror humidity sensor, *40–41*
Chiller optimization, 78, **226**
Coaxial cable, *see* cable
Colossus I, 53
Commissioning DDC, *216–218*
Compiler, 72, 83, *85–86*, **226**, 231
Computer-aided engineering (CAE), 63, *82–83*
Configuration, DDC, *81–88*
Conservation laws, energy and matter, 30
Construction Specification Institute, Manual of Practice, 208, 222
Controllers:
  application-specific, unitary, 1, 8, 43, 60, 65–66, 78, 81, 107, 127, **225**, **233**
  industrial, 7, 40, 47, 65, 224, 227, 232
Control loop tuning, 65–66, *101*, 223
Control relay, 6, 11, 46–48, 50, 200, 217, **227**, 230
Cornell University, xi, 4, 6, 96

CSA, *see* Canadian Standards Association
CSI, *see* Construction Specification Institute
CSMA, CSMA/CD, *see* carrier sense multiple access

D/A, *see* digital-to-analog
Daisy chain, 66, 137, **227**
DARPA, *see* Defense Advanced Research Projects Agency
da Vinci, Leonardo, 3
Data communication, 2, 6–8, 67, 73–75, 94, **107**, 108–193, 219, 223
Data communication equipment (DCE), **113**, 125–127, **227**
Data link layer, OSI, 112, 117, *128–129*, 133, 135–137, 139, 144, 146, 148, 154, 159, 174, 176–177, 182, 186–188, 191 192, 226–**227**, 230, 232–233
Datapoint corporation, 136, 225
Data terminal equipment (DTE), **113**, 116–117, *125–127*, 139, 141, 143, 151, **228**
DCE, *see* data communication equipment
DDC:
  commissioning, *216–218*
  configuring, *81–88*
  installing, *214–216*
  operation and maintenance, *219–221*
  specifying, *208–214*
  system design, *195–205*, 207–208
  training, *218–219*
Defense Advanced Research Projects Agency (DARPA), **227**, 233. *See* TCP/IP
DeltaNet Pascal, 84–85
Demand limiting, 161, **227**, 230, 233
Design, DDC system, *195–205*, 207–208
Digital, **227**. *See also* analog-to-digital, digital-to-analog
Digital Equipment Corporation (DEC), 73, 94, 153
  PDP-11, 73
  RSX-11M, 73
  VAX, 71, 73, 96
Digital-to-analog (D/A):
  conversion, *48*, 58, 66
  converter, **227**
DIN 43760, 17, 51, 210
Discriminator control, 77
Distributed processing, 1, 8, 94, 97, 109, 151, 154, 161, **227–228**
Doppler effect, 37–38
Downloading, 47, 74, 82, 87, 96, 129, 149, 165, 182, 192, **228**
Drebbel, Cornelius, 3
DSC-8500, 85. *See* CAL1
DTE, *see* data terminal equipment
Duty cycling, 76, **228**, 230

EBCDIC, 148, 169, **228**
Economizer control, dry bulb, *77*, 199, 202
Edison, Thomas, 3
EIA, *see* Electronic Industries Association
Electronic Industries Assoiation (EIA):
  TR-41.8, engineering committee, 81
  EIA-232, *Interface Between Data Terminal Equipment and Data Communication Equipment Employing Serial Binary Data Interchange*, 93–94, 123, *125–127*, 129, 151, 186, 191, 221
  EIA-485, *Standard for Electrical Characteristics of Generators and Receivers for Use in Balanced Digital Multipoint Systems*, 124, *127–129*, 151, 174, 182–183
  EIA/TIA-568-91, *Commercial Building Telecommunications Wiring Standard*, 81, 89
  EIA/TIA-569-90, *Commercial Building Standard for Telecommunications Pathways & Spaces*, 81, 89
EMCS, *see* energy management and control system
Energy management and control system (EMCS), 104, 191, 220, **228**
Energy management software, *76–78*
  boiler, chiller optimization, *78*, **226**
  discriminator control, *77*
  dry bulb economizer, *77*, **228**
  duty cycling, *76*, **228**
  enthalpy control, *77*, **228**
  load shedding, 65, *76–77*, **230**
  optimal start/stop, 47, 65, *78*, 161, **231**
  start/stop, *76*, 200
  temperature setback, *77*, 198–199, 202

Energy management software (*cont.*)
variable-speed pumping, 48, 76, *78*, 101, 124
VAV, 60, 65–66, *77–78*, 81, 225
ENIAC, 53
Enthalpy control, 77, **228**
Error detection, data communication, 6, 117, 136, 138, *144–146*, 186
checksum, *145*
CRC, *145–146*, 185, 188
parity, *145*
Ethernet, ISO 8802-3, 114, 121, 123, 129, 134, 136, 139–144, 147, 150, 174, 176, 182–183, 227, 230
Cheapernet, 140
StarLAN, 140
10BaseT, 140–141, 143
Expert systems, 80, *104*

Fairchild integrated circuits, 59
Faraday, James, 3
Feedback control, 1–3, 5, 9, 47–48, 50–51, 63–64, 66, 781 157, 201
Fiber optics, 45–46, 51, 67, 109, 118, 121, 134, 140–141, 150–151, 228, 232
cable, *121–124*
sensors, *45–46*
Fieldbus, *see* Profibus *and* SP50
Fire code, 118, 215
Fire control, *80–81*, 155, 161
Firmware, 54, 59, 63, 65–66, 81, 92–93, **228**
Flow diagram, design, *202–205*
Flow sensors, *29–39*
hot-wire anemometer, *34–36*
orifice plate, *32–34*
pitot tube, *31–32*, 34, 35
turbine, *36*
ultrasonic, *37–39*
venturi, *33–34*
vortex shedding, *36–37*
FORTRAN, 84, 160, **228**
FSP (facility-system-point) naming, 103
Full duplex, 116, 123, 148, 186, **228**
Fuzzy logic, *223*

Gateway, network, 161, 167, **229**
Gradual switch, 4–5
Graphical programming, 63, 83, 85, *87–88*, 93, 205, 224

Half duplex, 116, 126, 148, **229**
Hardware, DDC, 53–67, 225, **229**, 232
Harvard Mark I, 53
Honeywell, 4–6, 84
Deltanet Pascal, 84–85
Host computer, 85, 96, **229**
Hub, network, *131–132*, 137, *141*, **229**
Humidity sensors, *39–41*
thin-film polymer, *39–40*
chilled mirror, *40–41*

IAQ sensors, *40–41*
IBM, 6, 58, 60, 73, 82, 87, 94, 96, 132, 134, 169, 228, 233
PC, 60, 73, 82, 87, 96
System/7, 6, 73
Token-ring LAN, 132, 134
IC, *see* integrated circuit
IEEE, 72, 89, 133, 139, 143, 170, 216, **229**, 230
IEEE standards:
IEEE 587-80, *Guide for Surge Voltage in Low Voltage AC Power Circuits*, 216, 222
IEEE 754-1985, *IEEE Standard for Binary Floating-Point Arithmetic*, 170, 193
IEEE 802 LAN standards, *133–136, 139–145*, 147, 151, 182, 187, 192, 230
IEEE Standard 1003.1-1988, *Portable Operating System Interface for Computer Environments (POSIX)*, 72, 89
Input/Output (I/O) summary, 195–196, 198, *200–201*
Installing DDC, 214–216
Interactive computing, 73, 101, 158, **229**
Integrated circuit (IC), 13, 23, 43, 45, 54, 59, 60, 113, 137, 227, **229**, 233. *See also* large scale integration, very large scale integration
Integrated circuit temperature sensor, 23
Intel, microprocessors, 59–61, 67, 73, 127, 139

Intelligent buildings, 80–81
International Organization for Standardization (ISO), *see* ISO standards
Interpreter, source or pseudo code, 85–86, 92, **229**
Interrupt, hardware, 59, 70, 72, **229**
IP, Internet Protocol, *181–182*, 226–227, **229**, 233
I/P transducers, *48–50*
IPX, Novell Netware, *181–182*
ISA, 22, 42, 51, 67, 202, 205, 219
ISA standards:
- ISA S5.1, *Instrumentation Symbols and Identification*, 202, 205
- ISA S50.1-1982, *Compatibility of Analog Signals for Electronic Industrial Process Instruments*, 42, 51, 67

ISO standards:
- ISO 7498, *Open Systems Interconnection—Basic Reference Model*, 107, 151. *See* OSI
- ISO 8802-2, *Logical Link Control (LLC)*, 151, 187
- ISO 8802-3, *CSMA/CD*, 139, 151, 182, 192
- ISO 8824, *Abstract Syntax Notation One (ASN.1)*, 168, 193
- ISO 8825, *Abstract Syntax Notation One (ASN.1)—Basic Encoding Rules*, 169, 193
- ISO 8877, *Interface Connector and Contact Assignments for ISDN,* 141, 151

Johnson Controls, 4–5, 60, 84–86, 195

Knowledge-based engineering, 91

LADC, *116*
Ladder logic, 47, 232
LAN, *see* local area network
Large scale integration (LSI), 58, **229**. *See also* very large scale integration
Layers, OSI data communication, *see individual layers, i.e.,* physical layer, data link layer, network layer, transport layer, session layer, presentation layer, application layer. *See* also application program interface
Layers, OSI, mail analogy, 110–112
Line driver, 113, 116, 125, 186, 227, **229**
LLC, *see* ISO 8802–2
Load shedding, 65, 76–77, **230**
Local area network (LAN), 8, 114, 117, 120–121, 123, 125, 129, *133–141*, *146–148*, 151, 154, 159–160, 167–168, *174–186*, 221, 225–227, 229, **230**, 232–233
Local loop control, 80, 94, 199, **230**, 233
LVDT, 25, **230**

MAC address, 146, 176, 181, **230**
Machine language, **230**, 231
Maintenance scheduling, *81*
Manchester encoding, *114*, 140, **230**
Manufacturers, DDC, 235–236
Medium access control (MAC), 109, 133, *128–144*, 146, 168, 186, 227, 230
- master/slave, 7, 116, 128, *129–130*, 146, 174, 182–186, 230
- peer-to-peer, *128–130*, 183, 230
  - contention, 129
  - token-passing, 130

Memory, computer, 53–54, 59, 61–63, 69–70, 73–74, 86, 95, 165, 223–224, 226, 228–233. *See also* RAM and ROM
Metasys, 85, 88
Metering, hardware, 19, 24, 51
Metering, software, 69, 76, *79–80*, 98, 109, 207
Microcomputer, 7–8, 55, *58–60*, 228, **230**
Microcontroller, 54, *58–60*, 66, **230**
- Intel 8051, 60, 62
- Intel 8096, 60, 62
- Motorola MC68HC11, 60, 62

Microprocessor, *see* Intel, Motorola
Microsoft, 58, 71, 73
- DOS, 71, 73–*74*, 96
- Windows, 71, 73

Mode summary, 196, *200*, 202–203
Modems, **113**, 115–117, 123, 125–127, 129, 186–187, 191, 220–221, 227, **230**, 233
- Bell 202, 115–116

Modems (*cont.*)
 Bell 212, 115–117
Modulation, 113, *115–117*, 123–124
Monitoring, software, 79–80
Motorola, 60, *62*, 67, 73
 MC6502, 60
 MC6800, 60, 73
 MC680XX, 60, *62*, 67, 73
MS/TP, BACnet, *182–186*
Multiplexer, 7–8, *42*, 46, 54, 66, 229, **230**
Multitasking, 63, *70*, 96, **230**
Multivendor, *see* operator-machine interface, multivendor
MUX, *see* multiplexer

Network access method, *see* medium access control
Network layer, OSI, 111–112, *146–147*, 229, 232
Network layer, BACnet, 159, 174, *175–182*, 186, 226
Network topologies, *131–133*
Network-visible, making DDC devices, *154–155*, 158, 165–166, 230–231
Non-volatile memory, 228, **231**–232
NRZ signaling, *113–114*, 140, **231**
Numbering systems, binary, decimal, hexadecimal, octal, 54–58

Object, BACnet, *154–166*, 168, 171–172, 174, 190–192, **231**
Object code, 86, 225, **231**–232
OMI, *see* Operator-machine interface
Open Systems Interconnection (OSI), *see* layers, OSI data communication
 mail analogy, *110–112*
Operating system, 63, *69–74*, 86, 89, 92, 94, 96, 104, 143, 149, 228–229, 231
 DDC, *73–74*
 workstation, *73*
Operation and maintenance, DDC, *219–221*
Operator-machine interface, *91–104*
Operator-machine interface, multivendor, *101–103*
Optimizing functions, 8, 47, 65, 78, 161, 177, 226, 228, 230–**231**
Optimum start/stop, 47, 65, *78*, 161, **231**
OSI, *see* Open Systems Interconnection

Pascal programming language, 72, *84–85*, 160, 168–169, 232
PCI, protocol control information, 159–160, 170, 177, 179, **231**
PDU, protocol data unit, 159–160, 168–174, 177–179, 186–188, 190, **231**
Physical:
 layer, OSI, *112–129*, 133, 136, 139, 146, 148, 151, 154, 174, 176, 187, 191
 media, *117–128*. *see also* cable
PID control, 63, 84, 87, **231**
PIN diodes, *122*
Pixel, *95*, **231**, 233
PLC:
 see powerline carrier communication and programmable logic controller
PMI, **231**. *See* operator-machine interface
Polling, 129, 185–186, **231**
POSIX, 72, 89
Powerline carrier communication, *124–125*, **232**
Powers, Landis and Gyr, 5, 84
PPCL, 84–85
Presentation layer, OSI, 111, *148–149*, 169
Pressure sensors, *24–29*
 capacitive, *25*, 29
 inductive, *25–26*, 29
 piezoelectric, *27*, 29
 potentiometric, *27*, 29
 strain gage, *26–29*
Process control, industrial, 7, 40, 47, 51, 65, 67, 101, 202, 224, 227, 232
Profibus, 46
Programming DDC, *see* DDC, configuring
Programmable logic controller, 47, 65, 232
Protocol stack, 147–149, 169, 177, 225, **232**
Pseudo code, *86*, **232**
PTP, BACnet, *186–189*
Pulse width modulation, 48, *50–51*

RAM, 53, 59, 79, **232**, 233
Relays, *see* control relay

Repeater, LAN, 124, 140–143, *146–147*, **176**, 182, 186, **232**
Resistance, nominal, **14**, 16–18, 21, 23–24, 121
Resistance, temperature coefficient of, **14**, 19, 210
Reynold's number, 32
Ring topology, LAN, *131–134*, 136–138, 147, 185–186, **232**
ROM, 53, 59, 74, 150, 228, 231–**232**
Router, network, 135, *147–148*, 168, 176–182, **232**
RTD temperature sensors, 13, *14–23*, 42, 45, 200–201, 210, 221

SAMA PMC 22.1, *Functional Diagramming of Instrument and Control Systems,* 202
Schedule, temperature reset, *77*, 198, 207
Schedule, time-based, 8, 157, 172, 198–199, 207, 218, 227
Security control, *80*, 155
Security, data, 94, 111, 134, 187
Seebeck effect, *21–22*
Sensors, *see type, e.g.*, temperature
Sensors, general characteristics, *12–13*
Sensors, future, *45–46*
Session layer, OSI, 111, *148–149*
Sequence of operation, 195, 198–200, 205
Signal conditioning, 11–13, 17–18, 23–25, 27, *41–42*, 43, 46
Sintering, of thermistors, 19
Smart sensors and actuators, 45–46, 63, *66–67*
Software, DDC, *69–89*. *See also specific types, e.g.*, application software
Source code, 86, 92, **232**
Specifying DDC, *208–214*
SP50 committee, ISA, 46, *67*
Standards, *see standards-making organizations, i.e.*, ANSI, ASHRAE, DIN, EIA, IEEE, ISA, ISO, and SAMA
Star topology, LAN, *131–132*, 137, 140–143, 229, **232**
Strobe pulse, parallel communication, *113*, **232**
Supervisory control, 5, 70, 94, 228, **232**

TCP/IP, 227, 229, **233**. *See also* IP
Temperature Scale, International Practical, 13
Temperature sensors:
- integrated circuit, 13, *23*
- RTD, 13, *14–23*, 42, 45, 200–201, 210, 221
- thermistor, 13, *18–21*, 23–24
- thermocouple, 5, 13, *21–24*

Template programming, *82–83*, 87
Test equipment, DDC, 220–221
Texas Instruments, 59
Thermistor temperature sensor, 13, *18–21*, 23–24
Thermocouple temperature sensor, 5, 13, *21–24*
Thermostat, 4, 8, 48
Token-passing, 130. *See also* medium access control, and MS/TP
Top-down design method, *195–205*
T/P, *see* token-passing
Training, DDC, *218–219*
Tranducers, *11–44*
Translator, source code, 74, 83, *85–86*, 229, **233**
Transport layer, OSI, 111, *148*
Trend-log, 98, 220, **233**
Tuning, control loop, 65–66, *101*, 223
Twisted pair, *see* cable

UART, *113–114*, **233**
UDP, 181–182, **233**. *See also* IP and TCP/IP
Unicode character set, 58
Unitary controller, 8, 43, 60, *65–66*, 107, 127, **233**
Unix, 71–73
Uploading, **233**. *See* downloading
USART, 127, **233**
Utility software:
- DDC, *75–76*
- workstation, *74–75*

V/P transducer, *48–50*
Variable air volume (VAV), 60, 65–66, *77–78,* 81, 225

Variable speed drives, pumping, 48, 76, *78*, 101, 124
VAV, *see* variable air volume
Velocity sensors, *see* flow sensors
Ventilation, demand-controlled, *41*, 51
Very large scale integration (VLSI), 45, 54, 58, **233**
Volatile memory, 59, 232–**233**

Watt, James, 3
Wheatstone bridge, *16–18*, 26–27
Wide area networks (WAN), 148, **233**
Workstations, *see* operator-machine interface

XON, XOFF flow control, 181, 187–189, **233**

Zilog Z80, 60, 73